TURNER AND THE SCIENTISTS

TURNER AND THE SCIENTISTS

James Hamilton

Tate Gallery Publishing

Sponsored by Magnox Electric

LENDERS

The University of Bristol (cat. 90); The British Museum (cat. 36);
National Gallery of Scotland, Edinburgh (cat. 61); Edmund Fairfax-Lucy
(cats. 7, 8); King's Lynn Museums, Norfolk Museums Service (cat. 77);
Board of Trustees of the National Museums & Galleries on Merseyside
(Lady Lever Art Gallery, Port Sunlight) (cat. 86); University of Liverpool
Art Collections (cat. 100); National Gallery, London (cat. 89);
The Trustees of the Natural History Museum (cat. 3); Dr Jan Piggott
(cats. 108a, 108b); Preston Hall Museum, Stockton on Tees Museums
Service (cat. 53); Royal Institution of Great Britain (cats. 4, 5, 6, 109);
The President and Council of the Royal Society (cats. 1, 2);
The Board of Trustees of the Victoria and Albert Museum (cats. 68, 74);
Yale Center for British Art, Paul Mellon Collection (cats. 27, 80)

PHOTOGRAPHIC CREDITS

Beaverbrook Art Gallery, Fredericton; Birmingham Museums and
Art Gallery; Museum of Fine Arts, Boston; Bridgeman Art Library;
University of Bristol; British Museum; Edmund Fairfax-Lucy;
King's Lynn Museums Collection; University of Liverpool Art Gallery &
Collections; John Mills (Photography) Ltd.; National Gallery, London;
National Gallery of Art, Washington; National Gallery of Ireland;
National Museums & Galleries on Merseyside; National Portrait Gallery;
National Trust Photographic Library / John Hammond;
Natural History Museum Picture Library; Preston Hall Museum,
Stockton on Tees Museums Service; Erik Gould, Museum of Art,
Rhode Island School of Design; Royal Institution of Great Britain;
Royal Society; National Gallery of Scotland, Edinburgh; Tate Gallery
Photographic Department; V&A Picture Library; John Webb;
Richard Caspole, Yale Center for British Art

For my parents

front cover: *The Hero of a Hundred Fights* c.1806–7, reworked and
exh. 1847 (cat. 95), detail

frontispiece: *Forge Scene* 1796–7 (cat.41), detail

Published by order of the Trustees 1998
for the exhibition at the Tate Gallery
3 March – 21 June 1998

Published by Tate Gallery Publishing Limited
Millbank, London SWIP 4RG

ISBN 1 85437 255 6
A catalogue record for this publication is available from the British
Library

Designed and typeset by James Shurmer
Printed and bound in Great Britain by Balding and Mansell, Norwich

CONTENTS

Sponsor's Foreword   *page* 6

Foreword   7

Acknowledgments   8

Introduction   9

1 People and Ideas   12

2 Foundations: 'Consider the pleasure of being your   21
   own architect'

3 The Old Technology: Early Experience of Industry   37
   1775–1815

4 Observing the Sky: Meteorology, Astronomy and Visions   58

5 From Sail to Steam: The Absence of Trouble   74

6 Industry and Construction after Waterloo 1815–51   92

7 The Living Earth: Geology and Magnetism   115

   Notes   129

   Abbreviations   136

   Catalogue   137

   Index   142

SPONSOR'S FOREWORD

Magnox Electric has supported the Tate Gallery as a sponsor since 1992, during which time the company has been involved with a series of Turner events.

Our relationship with the Gallery and the Turner Bequest was established by our former Chairman, the late John Collier, a knowledgable and enthusiastic admirer of Turner's work. Beginning with sponsorship of the final exhibition in the Turner *Decade* series, we have also supported *The Essential Turner* exhibition and the Turner Scholarships.

Turner and the Scientists is the last exhibition in Magnox Electric's programme of sponsorship at the Tate and its special relevance to our business provides a fitting close. As a nuclear power generator, the origins of electricity production are rooted in the scientific discoveries of the nineteenth century, and the scientists whom Turner counted among his friends and acquaintances.

We share a particular interest in one scientist, who not only influenced Turner but continues to influence the lives of millions of people in the twentieth century. Michael Faraday's breakthrough discovery of 1831, which established the principle of electromagnetic induction, remains the scientific basis for generating electricity to this day.

We have been privileged to be associated with the Tate and hope that you enjoy *Turner and the Scientists*.

DENNIS JOYNSON
Managing Director
Magnox Electric

FOREWORD

Turner is well known as an artist who constantly looked ahead to new techniques and means of representation. However, it is perhaps less well known that his interest in the new extended to the great advances being made in science and technology during his lifetime. Turner formed friendships with some of the leading scientists, including Michael Faraday, Richard Owen and Mary Somerville. For this exhibition James Hamilton has looked into the ways in which Turner's intellectual interests and social contacts were to influence his subject matter, with many of his works representing his response to the technological developments of the time.

The Turner Scholarships were established in 1988 to fund original research into the works contained in the Turner Bequest at the Tate Gallery. The Scholarships provide an opportunity for scholars to research into a particular aspect of Turner's work, culminating in the publication of a catalogue and an exhibition. Since 1995 the scheme has been sponsored by Magnox Electric plc. We are extremely grateful for their continuing support, without which the opportunities for new research into unexplored aspects of Turner's work would be severely limited.

In 1995 the Scholarship was awarded to James Hamilton, University Curator at the University of Birmingham and the author of the recent widely acclaimed biography *Turner: A Life* (Hodder and Stoughton, 1997). We are particularly grateful to him for taking on the task of researching this far-reaching subject and for everything he has done in connection with the exhibition and catalogue.

Many of the works in the exhibition come from the Turner Bequest. However, we are grateful to a number of collectors and institutions who have made it possible not only to illustrate the people and their discoveries that were to have such an impact on Turner, but also a number of key works demonstrating his response. We would like to offer our sincere thanks to all those who have lent to the exhibition.

Nicholas Serota

ACKNOWLEDGMENTS

It is always a privilege and delight to work with the Turner Bequest, which is available to all in the Clore Gallery Study Room at the Tate. However, through the generosity of Nuclear Electric plc (latterly Magnox Electric plc) who financed this Turner Scholarship inspired by their visionary Chairman, the late John Collier, I have been able to work there for days on end, and to travel in pursuit of Turner.

The staff of the Tate, in particular David Brown and the Keeper of the British Collection Andrew Wilton, have been unfailingly helpful and generous with their time and expert counsel, and I would like to thank them and the Director, Nicholas Serota, and his colleagues Ann Chumbley, Andy Loakes, Helen Sainsbury, Ruth Rattenbury, Sarah Taft, Joyce Townsend, Ian Warrell and the former Keeper of the British Collection, Martin Butlin. The staff of Tate Gallery Publishing have guided this publication into the light, while my fellow Turner Scholars Peter Bower and Eric Shanes helped to guide me nearer to the light than perhaps I might have been without them.

Staff from institutions in the United Kingdom and the United States have been unfailingly helpful: Susan Bennett, Archivist at the Royal Society of Arts, London; Neil Brown and Wendy Sheridan, the Science Museum, London; Diana Chardin, Trinity College Library, Cambridge; Anne Cowne, Lloyd's Register of Shipping; Sarah Dodgson, Librarian, The Athenaeum; Stephen Eeley, Oxford University Museum; Frances Gandy, Girton College, Cambridge; Vincent Giroud, Beinecke Library, Yale University; Colin Harris, Bodleian Library, Oxford; Paul Joyner and Graham Jones, National Library of Wales; A.W. Nayler, Royal Aeronautical Society; Roger Penhallurick, Royal Cornwall Museum; Colin Penman and Nicholas Savage, Royal Academy Library; Margaret Penston, Royal Greenwich Observatory; Derek Phillips, Cyfarthfa Castle Museum; Lisa Spurrier, Berkshire Record Office; Adrienne Tooke and Pauline Adams, Somerville College, Oxford; J.M. Wraight, Admiralty Librarian.

Lenders to the exhibition have been generous and understanding in the administration of their loans. I would like to thank Patrick McCaughey, Elisabeth Fairman, Patrick Noon and Scott Wilcox, Yale Center for British Art, New Haven; Ann Compton, University of Liverpool; Irena McCabe and Dr Frank James, Royal Institution; the staff of the Library of the Royal Society; Michael Richardson, University of Bristol; John Thackray, Natural History Museum; Susan Lambert, Victoria and Albert Museum; and staff both in the British Museum and in all the other institutions who kindly agreed to lend works. I am particularly grateful to the private lenders Edmund Fairfax-Lucy and Dr Jan Piggott.

My colleagues at the University of Birmingham, Carl Chinn, Desmond Costa, Paul Smith, Mark Storey and John Thornes have read parts of my text and advised constructively upon it, as did John Gage, Evelyn Joll and William S. Rodner who all read it in full. Professor Rodner's book, *J.M.W. Turner: Romantic Painter of the Industrial Revolution* (University of California Press, 1997), was published just as this catalogue went to press. The author kindly sent me proofs at a late stage, and whatever final polish this text may have is in some way due to the quality of light that Professor Rodner has thrown onto the subject in his work.

During the course of the research I have had warm hospitality from my cousins Alistair and Erin Smith and family in Madison, Connecticut; the Master and Fellows of Trinity College, Cambridge; Professor Gillian Beer, Dr John Gage and Rowland Eustace. David Holmes, Registrar of the University of Birmingham, gave me leave of absence to work on the Scholarship. I thank them all, and in particular I thank my wife Kate and my children who took Mr Turner in for far longer than they had originally bargained.

INTRODUCTION

The sixty or more years of Turner's working life spanned the period of the most profound social, economic and scientific change that the world had ever seen, and Turner was its witness. By the 1790s England's main industrial towns were linked by a network of canals and navigable waterways. These carried industrial products slowly but directly to and from the centres of population. By 1851, the year of Turner's death, canals were challenged by the spreading railway system, roads had improved beyond measure, and in addition to industrial goods, people could now travel around Britain with more freedom and to greater effect than ever before. Moreover, for thirty-six years since the end of the Napoleonic Wars, Britain had been at peace with the world, if not with herself. In the last months of his life Turner visited the construction site of the Crystal Palace in London, the building which came to house the largest and most influential exhibition of the world's industrial products that had yet been seen.[1]

Turner was not unique in his witness of this span of events, but he was in the depth, nature and development of his expression of them. Rural and urban industrial activity seized Turner's attention from an early age. Throughout his career this fascination with the way things worked, or were made, or the people who made them, or the *effect* that scientific and technological advance had on the world and society around him, never really deserted Turner, even though it was from time to time submerged by other concerns. In the 1790s there was little, beyond his manifest physical and intellectual energy, that could distinguish Turner from other artists of his own age. He saw a watermill in a valley or an industrial landscape with the eye of an artist of the Picturesque movement, working with an awareness of fashion and the need to satisfy patrons, to gain a reputation and make a respectable income through his art (figs. 30, 38, 39).

But when Turner entered a forge for his own purposes, as he did in the late 1790s to make studies such as *An Iron Foundry* or the *Wilson* sketchbook studies (figs. 34, 40), he did so to observe and learn, and to reach a practical understanding of the appearance of working machinery. His knowledge developed to the extent that at some point in the first decade of the nineteenth century he was sufficiently aware of his subject to make a considered oil painting of the interior of a forge, complete with a system of toothed wheels and hoists. This was never exhibited at the Royal Academy,[2] and remained gathering dust in one of Turner's studio stacks until, at a Varnishing Day before the Academy exhibition of 1847,

fig. 1 Detail from
Dido and Aeneas
exh.1814
oil on canvas
(N00494)

the artist himself obliterated nearly half its surface and transformed it into a fiery allegory of mid-nineteenth-century industrial achievement, and of British – specifically and determinedly British – military glory. This was his tribute to the Duke of Wellington, *The Hero of a Hundred Fights* (fig. 115). The ability to pick up a subject again after more than forty years, clear evidence of Turner's latent knowledge of himself and his life's work, indicates the potency that he saw in technological subject matter, and recommends it as a distinct area of study.

While technology is the direct subject of figs. 34, 40 and 41, and a vehicle for allegory in *The Hero of a Hundred Fights*, it plays a subordinate, but significant, role even in such a historical narrative as *Dido and Aeneas* (exh. 1814; fig. 1). In the distance, amid the buildings of Carthage gleaming in the morning light, is an extensive view of Dido's busy shipyards where, we might suppose from Virgil's text, Aeneas's storm-damaged fleet is already being repaired. Classical subject matter was never, for Turner, so far removed from reality that it could escape the earthbound touch of technology as appropriate.

Turner had an irrepressible love of gadgets, a domesticated and charming expression of his personal engagement with technology. He carried an umbrella with a dagger concealed in the handle;[3] he used a painting table on a central column which he could swing to reach the colour he required;[4] he fitted a water closet into his rebuilt house in Queen Anne Street (see Chapter 2); and he loved the paraphernalia in John Mayall's photographic studio (see Chapter 1). Added to this, Turner had a practical interest in the chemistry of colour making, but one driven by his wish to save money by doing it himself rather than by a need to conduct chemical experiments: his *Chemistry and 'Apuleia'* sketchbook[5] contains pages of recipes gleaned either from published sources or from instruction from tradesmen who prepared colours for artists.

This publication aims to reveal the impact that technological and scientific subjects and ideas had on the development of Turner's art from its earliest days, and its extent within his subject matter. It will examine, as far as can be determined, the nature and depth of his associations with scientists, and the influence they had on the direction of his thought. In looking at Turner's work from a hitherto unfamiliar angle, it will illuminate formerly unresolved aspects of his character and achievement, and perhaps correct imbalances. Turner's greatest champion since 1843, the year the first volume of *Modern Painters* was published, has been John Ruskin (1819–1900). The words of Ruskin, who shared with Turner a real enthusiasm for geology (see Chapter 7), have been followed for generations, perhaps too closely, as the lodestone for students of Turner. Ruskin's writings have directed the study of the subject and have established the syntax of its language; but we should remember that magnetic north is not true north and that, although Turner was tremendously important to Ruskin, Ruskin had limited importance for Turner.

During the course of its evolution the exhibition that this catalogue is published to accompany has had its share of title changes. It began as 'Turner and Technology', a title which would reflect a concern with subject matter and practical involvement. Although 'Turner and Architecture, Science and Technology' might have been clearer, it has ended as 'Turner and the Scientists', which suggests a wider emphasis on people and ideas.

Turner moved like a cat among the artistic and scientific society of his day. There are many clues to where he went, but although his meetings with, for example, Mary Somerville and Richard Owen were extensive and recurring, we know little or nothing of what was said. He was not alone in his friendships with scientists, for in the same social circles from the 1810s were many of his artist friends, Augustus Wall Callcott, Francis Chantrey, Charles Eastlake, Thomas Lawrence, Thomas Phillips, and gatherings with scientists were as much part of their social calendar as his. In his mature years Turner's response to scientific and technological subjects was oblique and absorbing, the product of one of the richest inner lives that any human has expressed.

1

PEOPLE AND IDEAS

Somerset House in the Strand, designed by Sir William Chambers, was the home of the Royal Academy, the Royal Society and the Society of Antiquaries. These three learned bodies moved there in 1781, and made it the nation's centre of intellectual exchange and debate. One immediate benefit of this proximity was that members of one society could conveniently attend meetings of another; indeed some were members of two or all three. While their aims and intentions diverged, each society thrived on the support of the limited number of intellectuals, dilettanti and members of the aristocracy who had social and financial connections in town and country.

The Royal Society, as its charter of 1663 defined, took responsibility 'for improving naturall Knowledge', or 'Natural Philosophy' as science was then called. It held regular meetings, weekly during the season, at which papers were given to members and their guests on subjects ranging widely across the natural world. The Society of Antiquaries, a body of historians and archaeologists, met to discuss issues of the recent and distant past, and to exchange objects and information. At the Royal Academy teaching was no less central but took a different form. While the topics at the Royal Society and the Antiquaries changed meeting by meeting, at the Academy the professors presented series of lectures annually to students of painting, sculpture and architecture. Though the content of their lectures evolved, the Academy professors repeated their teaching to new students year by year.

The Journal Books of the Royal Society record the names of guests introduced to meetings, and these reveal something of the flow of artists and their putative scientific interests. The watercolourists Henry Edridge and Thomas Hearne were regular attenders in the 1790s, being introduced usually by their patron Sir Henry Englefield (1752–1822), President of the Society of Antiquaries. The architects John Soane and Henry Holland attended together in 1795 to hear Dr George

Pearson's account of a 'new kind of steel called Woots at Bombay', while the three watercolour painters John Laporte, John Varley and John Jackson heard the chemist Charles Hatchett give his 'Analysis of the earthy substance from New South Wales, called Sydneia'.[1] There are reports of artists attending meetings that were of direct relevance to their profession: when Thomas and William Daniell, Revd Robert Nixon, 'Mr Smirke' and Sir William Beechey attended in February 1801, they heard Count Jacques-Louis Bournon (1751–1825) speak on 'the Arseniates of copper and iron discovered in Cornwall'. These minerals, used in the manufacture of pigments, were described by the speaker as 'beautiful deep blue or fine grass green or bluish green, very beautiful'.[2] And when Sir Benjamin Thompson, Count Rumford, gave his paper on 'Methods of measuring the comparative intensities of light emitted by luminous bodies' in 1794, he ended by expressing a desire for further investigations into the subject, recommending it 'to the contemplation of Philosophers, Opticians and Painters'.[3] This paper was attended by 'Mr Turner', introduced by the surgeon William Blizard FRS. It is one of many instances of 'Mr Turner' appearing in the lists, and, tantalising though it certainly is, we cannot categorically say that any of them is J.M.W. Turner.

The rooms of the Royal Society were entered to the left of the columned portico off the Strand, while the front door of the Royal Academy was to the right. Thus, for the fifty-six years 1781 to 1837, when the Academy moved to Trafalgar Square, only about thirty feet separated these two learned teaching bodies. That narrow gap has been steadily widening ever since.

Another organisation which aimed to promote developments in the practical applications of art and natural philosophy was the Society for the Encouragement of Arts, Manufactures and Commerce. This had been founded in 1754 to stimulate invention and ingenuity in Britain and the colonies, and to act as a conduit

fig. 2 Thomas Phillips, *3rd Earl of Egremont* oil on canvas *Petworth House*

eighteen and aware of the social, financial and artistic importance of being acknowledged by the establishment, Turner entered a Premium competition run by the Society of Arts for drawing landscape, and won the coveted Silver Palette[6] (see Chapter 3).

A new organisation whose aim evolved to promote the wider understanding of natural philosophy came into being in London in 1799. This was the Royal Institution, founded through the initiative of Sir Benjamin Thompson, Count Rumford (1753–1814) as an establishment 'for diffusing the Knowledge, and facilitating the General Introduction of Useful Mechanical Inventions and Improvements; and for teaching, by courses of Philosophical Lectures and Experiments, the application of Science to the common Purposes of Life.'[7] Rumford became the first Secretary to the Institution

directing funds and patronage. Its guiding ideal was to seek means of enlarging the nation's productive capacity so that Britain would prosper and exceed France in world influence. In the words of the society's founder, William Shipley, the aim was 'to render Great Britain the School of instruction as it is also the centre of traffic to the greatest parts of the known world'.[4]

The home of the society, opposite the Adelphi in John Adam Street, was about five hundred yards along the Strand from Somerset House.[5] Until he was twenty-five, Turner himself lived and worked in Maiden Lane, Covent Garden, two hundred yards north of the Strand. Thus, at the beginning of his career he was placed geographically at the heart of the British cultural establishment, the point where art and natural philosophy came together as never before or since. In 1793, aged

fig. 3 Sir Francis Chantrey, *Bust of John Fuller MP* 1820 (cat. 4)

fig. 4 Thomas Phillips, *Sir Humphry Davy, Bt* 1821 oil on canvas *National Portrait Gallery*

and, drawing aristocracy and industrialists into his orbit, encouraged early support from Lord Egremont (1751–1837; fig. 2) and the Sussex MP and landowner John Fuller (*c*.1756–1834; fig. 3).[8] The Institution's philanthropic intentions were clear, as were its practical concerns for economy in wartime, and its determination to be 'a grand repository of all kinds of useful mechanical inventions', with a laboratory and lecture theatre for 'teaching the applications of science to the useful purposes of Life'. It did nonetheless reflect a broad range of general public educational interests, promoting lectures by painters including John Constable and Thomas Phillips, poets such as Thomas Campbell, and inventor-entrepreneurs such as Captain George Manby. In founding the Royal Institution, Rumford identified a new social need which had evolved separately from the theoretical researches of the Royal Society, the practical art teaching and exhibitions of the Royal Academy and the commercial aspirations of the Society of Arts.

The chemist Humphry Davy (1778–1829; fig. 4) was appointed resident Lecturer of the Royal Institution in 1801. He was a charismatic speaker, and came to London from youthful successes in Cornwall and Bristol. Davy carried out experiments in the chemical laboratory, and lectured to large and enthusiastic audiences of men and women from all classes of society on chemistry, geology, electricity, galvanism and other subjects, always with practical reference to the application of science to agriculture, domestic industry and manufacture, on which modern life was rapidly beginning to depend. He gave practical and spectacular demonstrations to explain his points, during a lecture in 1811 setting off a model volcano to 'tumultuous applause and continued cheering of his audience'.[9]

The explosion of Davy's volcano, probably by the action of water on potassium permanganate, is a useful metaphor for the explosion of public interest in natural philosophy in the early nineteenth century. With widely advertised prizes available for ingenious ideas made practical, and for solutions to technical problems that had long aggravated British industry and commerce, opportunities grew for every man and woman to contribute to the growing welfare of Britain. In addition, natural philosophy societies began to appear in London and provincial towns ensuring through lectures and demonstrations that those who wanted to learn about the new ideas could do so reasonably easily.[10]

The key issues explored by pioneer scientists – I shall call them scientists from now on for convenience, though the word did not come into general use until the 1830s – included chemistry, cautiously defined by Davy in 1800 as 'in its present state … simply a partial history of phenomena, consisting of many series more or less extensive of accurately connected facts'.[11] Chemistry was then an almost boundless field, encompassing areas that have since been closed off from it and colonised as physics, including the identification and isolation of elements, the discovery of new chemical compounds, the nature of light and heat, the nature of electricity and its relationship with magnetism, and the processes of volcanic eruption. Vulcanism linked chemistry to geology, a science also explored by the omnivorous genius of Davy and by small armies of peripatetic men with rock-hammers, including Charles Lyell (1797–1875), William Buckland (1784–1856; fig. 5), John MacCulloch (1773–1835), Adam Sedgwick (1785–1873) and William Whewell (1794–1866). Geologists then, as now, rumi-

fig. 5 Thomas Phillips, *William Buckland* 1831 oil on canvas
National Portrait Gallery, London

fig. 6 Antoine Claudet, *Charles Babbage* 1851 daguerreotype
National Portrait Gallery, London

nated upon the nature and composition of the earth, and argued passionately about its creation, its age, its history as reflected in the fossil record, and the evidence of the biblical flood, one of the greatest controversies of the age.

The study of fossils brought naturalists to geology, and away again to develop the parameters of their own particular science, natural history. This divided itself into two main groupings at least: those who studied creatures, zoologists, and the botanists. The former, whose research had grown in its turn out of the disciplines of medicine and anatomy, included William Broderip (1798–1859), Sir Anthony Carlisle (1768–1840) and Sir Richard Owen (1804–92). The leading botanists of the day included Sir Joseph Banks (1743–1820) and Sir William Hooker (1785–1865).

As well as looking down at the earth, scientists looked out to the stars, developing the ancient and elemental science of astronomy. Here Sir William Herschel (1738–1822) had been pre-eminent in the latter decades of the eighteenth century, developing telescopy to an unprecedented technical precision, enabling him to discover stars and the planet Uranus, and to make new maps of the heavens. He passed his insight to his son John, who carried the work further, mapping the stars of the Southern Hemisphere, extending the frontiers of mathematics, and latterly pioneering the development of photography.

A pioneer of optics was the Scotsman Sir David Brewster (1781–1868). Among his many achievements was the invention of the kaleidoscope (*c.*1814), which resulted from his work on the laws of the polarisation of light. As editor of the *Edinburgh Encyclopaedia* (published from 1830), Brewster was in close touch with scientists and other savants all over Europe. His 'queer discoveries on the spectrum'[12] fascinated his colleagues and excited Turner's interest – see Chapter 4 below.

John Herschel (1792–1871), who was knighted in 1831, was part of the close-knit circle of friendships which was a natural consequence of scientific discovery, and which included the mathematicians Charles Babbage (1792–1871; fig. 6) and Mary Somerville (1780–1872). Babbage, a man as gregarious and omnivorous as Humphry Davy, moved mathematics towards the twentieth century, devising two extraordinary calculating machines, the Difference and the Analytical Engines which prefigured the computer. Mary Somerville (fig. 7), though living in Italy for much of her later productive life, wrote fluently and plainly, bringing the fruits of her research into magnetism, astronomy and geology to the widest possible audience of lay men and women through a series of lucid explanatory texts, *On the Connexion of the Physical Sciences* (1834) and *Physical Geography* (1848). The former, which ran to many editions, established Somerville as a pioneer of practical sciences and their dissemination, foreshadowing in the twentieth century Julian Huxley, David Attenborough, Stephen Hawking and Richard Dawkins. Through her extended periods on the Continent, Somerville fostered new webs of friendship with European scientists, and acted as one of a number of channels through which their ideas and discoveries could be communicated to their British colleagues. Mary Somerville's friend, the writer Maria Edgeworth, wrote of her: 'Mrs Somerville is the lady who, La Place says, is the only woman in England who

fig. 8 Matthew Noble, *Bust of Michael Faraday* 1854 (cat. 2)

fig. 7 Sir Francis Chantrey, *Bust of Mary Somerville* 1840 (cat. 1)

understands his works. She draws beautifully; and while her head is among the Stars, her feet are firm upon the earth.'[13]

The giant among nineteenth-century giants was Michael Faraday (1791–1867; fig. 8), a devout Christian of the Sandemanian sect who learnt his science as Humphry Davy's assistant at the Royal Institution and travelled with him on the Continent. In 1827 Faraday succeeded Davy as Professor of Chemistry at the Royal Institution, where his precision of thought and hand, his patience and percipience, set new standards in laboratory practice. Through lengthy and persistent experiment Faraday demonstrated the relationship between electricity and magnetism which led directly to the development of modern techniques of electricity generation and storage, and ultimately to its distribution as a public utility.

The sciences that were being researched and widely discussed in Turner's maturity were perceived as an interconnected web: chemistry, geology, natural history, astronomy and mathematics depended closely one upon the other, as scientists themselves exchanged ideas across their none-too-rigid boundaries. Mary Somerville outlined the nature of the continuum in the introduction to *On the Connexion of the Physical Sciences:*

Astronomy affords the most extensive example of the connection of the physical sciences. In it are combined the sciences of number and quantity, of rest and motion. In it we perceive the operation of a force which is mixed up with everything that exists in the heavens or on earth; which pervades every atom, rules the motions of animate and inanimate beings, and is as sensible in the descent of a rain-drop as in the falls of Niagara; in the weight of the air, as in the periods of the moon.[14]

The sciences also fostered their connections and advances through societies, museums and gentlemen's clubs. In addition to the five leading organisations described above, societies devoted to particular disci-

fig. 10 Michael Wagmüller, *Bust of Sir Richard Owen* 1871 (cat. 3)

plines grew up – the Geological Society (founded 1807), the Analytical Society (1812), and the Zoological Society (1826). These were places for social gathering as well as scientific discussion, as were the homes of scientists and others on the social fringe of the subjects. Another important gathering place was the Athenaeum Club (1824), of which Turner was a founder member. Although there may have been other routes by which Turner met his scientist contemporaries, it was through breakfast parties, soirées and other social gatherings during his mature years that we know he had many fruitful and extensive friendships with scientists.

During his 1819 visit to Rome, for example, Turner met Sir Humphry Davy and his wife, although he had almost certainly also known them in London, for they had friends in common, including Chantrey, Lawrence

fig. 9 Harriet Moore, *The Magnetic Laboratory of Michael Faraday* c.1850 (cat. 5)

fig. 11 Mary Somerville, *Mole de Gaeta* 1839 (cat. 7)

fig. 12 Mary Somerville, *Italian Landscape: View towards a Bay* 1830s (cat. 8)

fig.13 Charles Mottram after John Doyle, *Samuel Rogers at his Breakfast Table in 1815* c.1823 engraving *Tate Gallery*

and Wilkie. In the 1820s and 1830s, at *conversazioni* with the lithographer Charles Hullmandel (1789–1850), through his friend Harriet Moore (fig. 9), and perhaps by other routes, Turner came to know Michael Faraday and discussed with him the nature of pigments and of the effects of light in the sky.[15] Mary Somerville and her husband Dr William Somerville became Turner's firm friends, and Turner was particularly fascinated by Mary Somerville's experiment on the magnetising properties of violet light.[16] His further interest in magnetism, which would have been fostered by his friendships with Mary Somerville and Michael Faraday, is discussed below in Chapter 7. Turner was a fellow dinner guest with Charles Babbage[17] and attended some of Babbage's popular and crowded soirées in Dorset Street, Manchester Square, where demonstrations of the Difference Engine were a special attraction: 'It will make a cube, eat a

bit of its tail, and tabulate queer functions,' Babbage told Herschel.[18] Turner paid regular visits to the apartments of Sir Richard Owen (fig. 10) in the Royal College of Surgeons in the 1830s and 1840s,[19] and up to the last years of his life he attended soirées at the Royal Society – the last recorded took place in 1848 or 1849 during the Presidency of the astronomer the Earl of Rosse (PRS 1848–54).[20]

Turner's curiosity for the sciences was matched by a corresponding practical interest in art among scientists. Faraday's two albums of miscellaneous prints and drawings, which include two Rembrandt etchings, survive at the Royal Institution, as do his volumes of engraved and lithographed portraits of scientist colleagues. These, as a trained book-binder, he mounted and bound himself.[21] Babbage also collected portraits;[22] John Herschel drew with great skill with the aid of a *camera lucida*;[23] and Mary Somerville was a highly accomplished and dedicated painter (figs. 11,12), who wrote with a painter's eye:

In this wide expanse [at Lake Albano] we could see the thunderclouds forming and rising grandly over the sky before the storm began. In the early morning I have frequently watched the vapour condensed into a cloud as it rose into the cool air; in the evening the sunsets were glorious. Fascinated by the brilliant colouring I attempted to paint what Turner alone could have represented.[24]

Other social catalysts of the period included the banker, poet and art collector Samuel Rogers (1763–1855), who held regular breakfast parties at his house in St James's Place (fig. 13). Turner attended many of these and other gatherings at the homes in Thames Ditton of the surgeon James Carrick Moore (1762–1860)[25] and in Lincoln's Inn Fields of the architect Sir John Soane (1753–1837). Soane had turned his house into a museum of art, antiquities and neo-classical taste, and welcomed his friends and contemporaries to parties and gatherings. The greatest of these was a three-day affair in 1825, at which Turner was present, to welcome the arrival of an Egyptian coffin that he had acquired. The publisher John Murray II (1778–1843) and his son and successor John Murray III (1808–92) held parties at their house in Albemarle Street, as did the surgeon Thomas Pettigrew (1791–1865) nearby in Savile Street. Turner was a regular guest of all.

In the late 1840s Turner befriended the young daguerreotypist J.J.E. Mayall (1813–1901) who had a shop in the Strand.[26] He showed close interest in the chemistry of photography, discussed Mary Somerville's experiments with the magnetising effects of violet light and sat for several daguerreotype portraits.[27] Turner came again and again to the shop, apparently incognito, but this is hard to believe.[28] Though born in England, Mayall had spent some years working in the United States, and Turner did not tire of hearing Mayall talk about the wonders of the Niagara Falls. Turner's visits to Mayall's shop ceased abruptly in 1849 after he had been recognised, but there did nevertheless remain one vestigial connection between the painter and the photographer which may have been due directly to the

brief friendship. Daniel J. Pound, the son of Turner's last mistress Sophia Booth, became an accomplished wood engraver and worked for Mayall. Among the daguerreotypes that Pound engraved were portraits of scientists and others known to Turner, including Sir David Brewster.

Soon after he returned from his first visit to Italy early in 1820, Turner confirmed his decision to have his house and gallery gutted, reorientated and rebuilt so that its principal entrance was a new double-frontage in Queen Anne Street, rather than the less imposing entrance round the corner in Harley Street. From the way that he furnished and decorated the house with plentiful and good quality furniture and fittings,[29] we can draw the conclusion that he did at least have the intention of entertaining friends old and new, and returning the hospitality he was being widely given. Four plaster casts from the Parthenon frieze were mounted in the hall, and other fittings included a cast of Apollo, some marble sculpture pedestals, mahogany furniture, crested shield-back chairs, bright yellow curtains and perhaps yellow walls in the parlour – all clear indications that Turner wished to present himself as a mature gentleman of means, with classical tastes and interests. It seems that at this time also he sealed his letters with wax impressed with the classical profile of an old man's head.[30] These clear statements of personal style reflect a modern outlook which accords well with his sociability and the range of his friendships among artists, collectors and scientists. It seems, however, that events may have conspired against his entertaining much at home: his father, who lived with him, was already in his mid-seventies, and his housekeeper Hannah Danby may by now have developed the malignant facial skin condition that plagued her latter years. Referring to Turner's empty dining table, Samuel Rogers remarked: 'It was wonderful … but how much more wonderful it would be to see any of his friends around it.'[31] This, however, does suggest that Rogers at least saw Turner's dining table.

FOUNDATIONS: 'CONSIDER THE PLEASURE OF BEING YOUR OWN ARCHITECT'

The subject of architecture and construction is a leit-motif that runs throughout Turner's career. His early watercolours are predominately of architectural subjects, set to a greater or lesser extent in landscape, and from these careful studies there is a clear line of ascent to the glistening views of Venice or Carthage of Turner's maturity.

Turner was also a practising architect, having trained with some of the leading London architects of the late eighteenth century. His output was small, but it was neat, satisfying and thoroughly contemporary, and evidently gave him great pleasure to accomplish. It was through architecture that Turner gained his first knowledge of how things were designed and made. Though arguably marginal to the main theme of 'scientists', Turner's friendships with architects were founded on common professional and technical interests, and for this reason they find a place here.

From as early as 1790, when Turner was fifteen years old, he paid close attention to perspective techniques and practised them in his drawings. There is sufficient technical improvement between the group of juvenile views of Margate, drawn in the late 1780s,[1] and the rigorous Wanstead New Church (1789–90) (fig. 14) squared up with construction lines, to suggest that in the few years between them the young Turner began to be taught to draw in perspective, and to have access to books on the subject. There were many routes by which Turner might have developed this knowledge, which he also exercised in other studies such as those of Radley Hall and Newnham Courtney in the *Oxford* sketchbook of c.1789 (fig. 15), and in seeking them out he differed little from other aspiring young artists of his generation. The topographer and architectural draughtsman Thomas Malton (1748–1804) taught perspective drawing from his rooms in Conduit Street from 1783 to 1789, and it is reasonable to accept the report of Turner's first biographer, Walter Thornbury, that Malton was his 'real master'.[2] Other teachers to whom Turner had

access were Thomas Sandby (1721–98), Professor of Architecture at the Royal Academy, and his brother Paul (?1731–1809).

One of Turner's special advantages over his contemporaries, however, was that by the age of sixteen he was already well travelled in southern Britain. Though his home was in Covent Garden, he had lived for a year or more in Brentford and Sunningwell, near Oxford. He had travelled to Margate in the 1780s, and in 1791 made his first journey to Bristol. In Brentford, where, from 1785 he lived at the centre of the new town with his uncle and aunt, he also began his lifelong friendship with the young Henry Scott Trimmer.[3] Trimmer's father, the brickmaker James Trimmer, was a colleague and undoubtedly also a supplier, of the Brentford mason-builder Thomas Hardwick the Elder. His mother, the pioneering educationalist Sarah Trimmer (1741–1810), was the daughter of Joshua Kirby, one-time clerk of works to the royal palace at Kew, who wrote the highly influential perspective teaching manual *Dr Brook Taylor's Method of Perspective Made Easy* (1754). This fortuante combination of circumstances could have given Turner early access to both the experience of practical construction and the techniques of perspective for artists. It can be no coincidence that, of all architects he might have been apprenticed to in London, the one who showed concern about his future and an awareness of his special talent was Thomas Hardwick of Brentford's son, Thomas Hardwick the Younger (1752–1829),[4] who designed St Mary the Virgin, Wanstead (completed 1790), the subject of fig. 14.

During Turner's youth London was alive with new construction. The Building Act had come into force in 1774, laying down minimum standards, reducing the risk of fire, creating squares and widening streets and terraces. For the first time, a sense of order was being imposed on the planning of London west of the City, and this had its effect all the way down the social scale from first- to fourth-rate housing. Within half a mile of

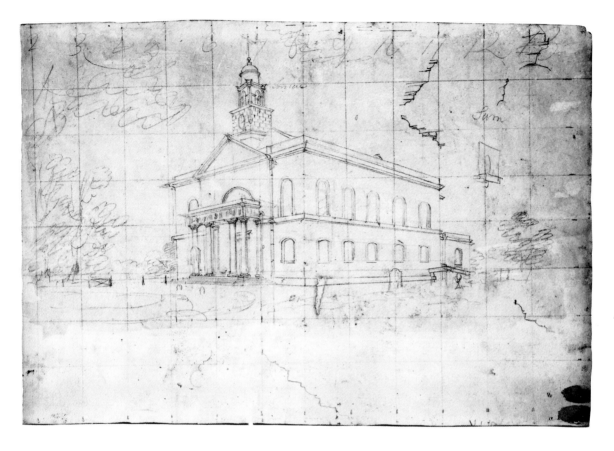

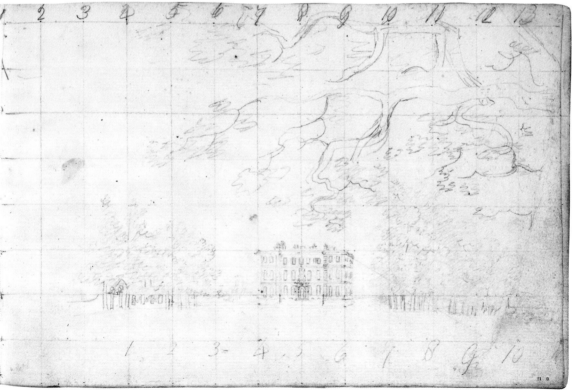

Maiden Lane there was much grand new architecture and building activity. The Adelphi off the Strand had been completed; Somerset House was being built to William Chambers's designs; the Prince of Wales was commissioning Henry Holland to enlarge Carlton House; and Thomas Hardwick was overseeing the rebuilding of Inigo Jones's St Paul's Church, Covent Garden. The acres flanking Oxford Street – the Portman, Portland, Grosvenor and Bedford estates – were being built during Turner's boyhood. To an eager and curious young boy London was one enormous extended building site, with many opportunities to meet and talk to architects and masons, to learn from observation about architectural style and ornament, and to practise drawing in perspective. It was this redoubling of London, perhaps even more than the activity of the river or the clutter and colour of Covent Garden Market, that gave such a special cast to Turner's boyhood, and underpinned his youthful enthusiasms for drawing architectural subjects.

With Thomas Hardwick's support, Turner became a probationer at the Academy in Somerset House in December 1789, and studied there assiduously by drawing from plaster casts before being promoted to the Life Class in 1792. Among the lectures he will have attended were those by the Professor of Architecture, Thomas Sandby, who covered the rise of architecture, the classical orders, the modes of building of other nations, plans and construction, and utility and beauty.[5] Here and there Turner left clues which suggest that he was attentive to the verbal language of architecture which Sandby and Hardwick would have taught him: he has written 'Ionic' beside the bell-tower in *Wanstead New Church*, and 'Tuscan / Doric / Corinth' on another sheet.[6]

Although the routine Academy teaching was directed towards developing a proficiency at drawing the figure, Turner's ambitions as an artist took him into wider fields. The writings of William Gilpin, Richard Payne Knight, Edmund Burke and Uvedale Price, all published in the second half of the eighteenth century, gave a structure to artistic theory and fuelled a taste for

(*left, top*) fig. 14 *Wanstead New Church* 1789–90 (cat.9)

(*left, bottom*) fig. 15 *Early Perspective Study: North Front of Radley Hall c.*1789, from *Oxford* sketchbook (cat. 18)

Picturesque landscape views. Turner was just one of the artists who supplied the late eighteenth-century market in authoritative drawings and engravings of buildings in their landscapes, travelling hundreds of miles in the 1790s to collect his material, west as far as Cardigan, north up to Berwick, east to Lincolnshire, and south to Kent, Sussex and the Isle of Wight. The calibre of the drawing in more than two hundred finished water-colours of architectural subjects made in the 1790s[7] shows how deeply Turner had absorbed the technicality of architectural construction. He could draw with such technical skill because he knew how buildings were made and, having watched them being built, he knew also the way they fell down and clattered into ruins. Particularly revealing are the unfinished drawings, those left abandoned for some reason at the pencil stage. He would use a ruler to lay in perspective lines and even to ensure that a tower was straight. Knowing his time on any particular subject was limited, he might draw half a gothic facade and leave the rest blank because the building was more or less symmetrical.[8] Such assurance allowed Turner to create extraordinarily daring, though not necessarily original, compositions – he mastered the highly complex geometry of the buttresses, windows and tower of St Mary Redcliffe, Bristol (1791–2), he created a partial vista of St Lawrence's Church, Evesham, beyond an architectural screen (*c.*1794)[9]; and there are many other examples. In looking at these and other very clever watercolours it is hard to avoid the conclusion that the young Turner may also, in fact, have been showing off.

Central among Turner's early series of architectural subjects are his views of Oxford, made from the late 1790s. He had hovered on the fringes of Oxford since his boyhood: there is a wistful, distant view of the town drawn in the late 1780s.[10] Later drawings, most of which were commissioned to be engraved for the annual Oxford *Almanack* depict the Oxford colleges one by one. As a collection these are the first tokens of Turner's life-long romance with learning and with the University of Oxford, a town where he might always be welcome but never became fully a part of. The unfinished *St Mary's and the Radcliffe Camera from Oriel Lane, Oxford* (mid 1790s; fig. 16) may have been an initial idea for an *Almanack* subject, even though its vertical composition suggests otherwise. The middle ground includes some scaffolding beside a building, evidence of the everyday, and part of the staffage of figures, dogs and carts that is so characteristic of Turner. In two other engraved *Almanack* subjects, the views of Worcester and Exeter Colleges, we see workmen digging the road or laying stones. This is not a metaphor that Turner ever over-emphasises, but in the activity of construction in these views we may see the artist speaking to us of building education. On the other hand, Oxford colleges are built of such soft stone that repairs are continuous and widespread.

One of Turner's many patrons in the 1790s was William Beckford (1759–1844), generally said to be 'England's wealthiest son'. From 1796 Beckford had been obsessed with the construction of Fonthill, a vast gothic pseudo-abbey overlooking Salisbury Plain designed by James Wyatt (1746–1813). While Fonthill was being designed, Turner was engaged by Wyatt to make a large presentation watercolour of the architect's ideas – part architect's impression, part artist's idealised romantic vision to impress the client (fig. 17). This was exhibited at the Royal Academy in 1798 as Wyatt's own work. On the strength of his growing reputation as a painter of sublime landscapes, as much as on his evident abilities as an architectural draughtsman, Turner was

(*right*) fig. 17 James Wyatt and J.M.W. Turner, *A Projected Design for Fonthill Abbey, Wiltshire* 1798 watercolour *Yale Center for British Art, Paul Mellon Collection*

(*left*) fig. 16 *St Mary's and the Radcliffe Camera from Oriel Lane, Oxford* mid 1790s (cat. 12)

himself invited by Beckford to Fonthill in 1799 to make a set of watercolour drawings of the building. Although Beckford received five finished views (exh. 1800),[12] each setting the abbey in the far distance, it is possible that these were not the compositions originally intended by either artist or commissioner.

There is in the Turner Bequest a further small group of unfinished pencil drawings taken from the dismembered *Fonthill* sketchbook (fig. 18 and cat. 11), which are quite different in kind and intention to the Fonthill watercolours. Their existence suggests that Beckford or Wyatt originally wanted close-up views of the abbey under construction. Challenged by the complicated perspectives and the fascinating relationships between the building's elements, Turner invested many hours of work in these sheets. In the view of part of the south front of Fonthill (fig. 18) the tower is built of temporary shuttering, with scaffolding, and men on ladders hauling up materials. Turner has evidently enjoyed creating the clever syncopated rhythm of the ground floor colonnade, where the perspective dictates the way the trac-

eried windows are obscured bit by bit by the columns in front of them. Expressing such delicious detail was all part of the fun of making drawings of this kind, and Turner's aborted Fonthill series is directly in character with many of his architectural pencil drawings of the 1790s and 1800s.

One is led to ask why do these Fonthill drawings exist, why were they not finished, and why are the delivered watercolours of such a different type? The answer, which we can only guess at, may reveal a difference of opinion between artist and commissioner, Beckford (or Wyatt) wanting views of the completed abbey, but Turner neither able nor willing to make drawings of a tower or spire as if it were complete. Thus, perhaps, it was finally agreed to do the long viewpoints with the detail of the tower fudged by the distance.

Turner's training in architects' offices must have been more thorough and practical than has generally been allowed, for he took upon himself the task of designing and overseeing the building of his own gallery in 1803–4. He altered the gallery twice in the following fifteen years, the second 'alteration' amounting to a complete rebuilding and reorientation of his house and gallery; and early in the 1810s he designed Sandycombe Lodge, his country house in Twickenham. Even in the last few years of his life he had the roof line of the house he shared with Mrs Booth in Chelsea amended to install a rooftop balcony so he could sit up there and watch the river and the sky. Furthermore, he was confident enough in his powers of building management to amass, through inheritance and purchase at auction, property all over London and far afield in Buckinghamshire and Kent.[13] Turner was no mean amateur architect. He

had mentors and close friends in the profession and, as a Royal Academician, was at the cutting edge of architectural thinking. He discussed architecture with colleagues and patrons, and the subject was, of course, the foundation of his annual series of lectures as Professor of Perspective at the Academy from 1811.

Although no large-scale architect's plans survive for any of his buildings, there are dozens of small sketches of possible layouts of Sandycombe Lodge in the *Sandycombe and Yorkshire* (fig. 19), *Windmill and Lock*, and *Woodcock Shooting* sketchbooks.[14] The vitality and kinship of the layouts reflect the clarity of Turner's thought, and

fig. 18 *Part of the South Front of Fonthill Abbey, the Tower in Process of Erection* ?1799, from dismembered *Fonthill* sketchbook (cat. 10)

fig. 19 *Sketches and Calculations for Sandycombe Lodge* c.1810, from the dismembered *Sandycombe and Yorkshire* sketchbook (cat. 13)

fig. 20 Archive photographs showing the entrance (*left*) and curved staircase and arched corridor (*right*) of Sandycombe Lodge, Twickenham

suggest that he knew what he wanted at an early stage. The spirit which unites the sketches is compactness and symmetry, qualities which are characteristic of the small villas and *cottages ornées* of the period.[15] Turner takes full advantage of the slope in the ground, which falls away from the front of the plot. This allows him to make the house appear from the public road to be a modest two-storeys with low wings, while from the garden it expands into a rather grander three-storeyed house. Such public modesty of profile, and private enlargement, parallel the character that Turner himself tended to project. Sandycombe Lodge, with its wide eaves reminiscent of Italian domestic architecture, is a subtle and thoughtful design. In the purity of its lines it owes much to John Soane, having passed through a Repton-like *cottage ornée* stage during its evolution.

Professional attention to the design of Sandycombe Lodge carried with it a careful attention to details of quantities and expenditure. Turner was prudent as well as prolific. The *Windmill and Lock* and *Sandycombe and Yorkshire* sketchbooks have pages of lists and sums written in the small, neat, legible hand that Turner could adopt when he was of a mind to do so. There are notes of the overall dimensions of the house:

40 feet long	20 feet high West side
20 Broad	30 feet high East side

and of the rooms:

2 Rooms	15 x 18
1 Room	15 x 20
Kitchen	15 x 12
Side room	15 x 8
Upper room	15 x 8
do.	15 x 8
Small do.	10 x 8 1 closet

100 80	100
	80
	8000[16]

There are lists of quantities of fittings required:

4 Chimney pieces
7 Best Doors – 2 got.
8 Common Doors.
12 Windows.
5 Attic windows – 2 got.
2 Pr. of F[rench?] Window Doors – outward doors'[17]

The tone of this list suggests that Turner had no intention of cutting corners of quality or price; nor does the ultimate appearance of the house – and its durability – show that he ever did. The *Windmill and Lock* sketchbook was in use around 1811, and by the following year, when Turner was using the *Sandycombe and Yorkshire* sketchbook, costs had started to come in as building work progressed. Thus we read:

400 – Purchase	
400 – Building	
100 – Grounds	
900	

and on the same page:

Walls	80	)
Chimneys	20	) Brickwork
Doorways	10	)
	110	
Roof	50	
Floors	50	
Slates	20	
Plastering	50	
Ornamental	50	
	320 [sic]	80 additional

These are neat estimates, showing how Turner hoped the costs would fall out, and how much he would have in hand. Against these estimates are shown two payments:

[£] 10	15791	17 Jan[y] 1812
10	7633	26 Feb[y] 1812[18]

The evidence of the surviving notes suggests that Turner wanted to define his social and professional status through the style and quality of his own country house. He certainly proposed to entertain his friends at Sandycombe, and did so on many documented occasions. The style he presented through the architecture was that of a well-to-do gentleman with classic tastes – the low eaves, the Diocletian window to the basement kitchen visible from the garden front, and the dentilation of decorative brickwork under the eaves and around the rounded corners of the two wings. Inside the house he designed a curving staircase (fig. 20) with a clear glass oval skylight above, decorated with a classical flower motif, a sculpture niche halfway up the stair, and in the ground floor rooms some neat fireplaces on

fig. 21 *Studies inside a Picture Gallery* 1808, from *Tabley No. 3* sketchbook (cat. 19)

classical lines. Such style is forward looking and, in 1812, still at the leading edge of the passion for the Grecian manner which superseded the eighteenth-century Pompeian style of the Adam brothers and the Roman classicism of Chambers and Hardwick.[19]

A further pair of buildings which Turner almost certainly had a hand in designing are the lodge gatehouses at Farnley Hall. Turner's close friendship with Walter Fawkes of Farnley is chronicled elsewhere,[20] and although there is little documentary evidence that Turner designed the gatehouses, circumstantial and stylistic evidence for it is strong.[21]

When Turner came to rebuild his house and gallery in Queen Anne Street nearly ten years later, his desire for the classic personal profile had not wavered. Nor indeed had his care over cash-flow, which he noted in the *Paris, Seine and Dieppe* sketchbook (cat. 24). Further, his concern for comfort and modernity led him to invest the considerable sum of £14 (around £700 at 1990s'

prices) on a water closet.[22] Upstairs, Turner's new gallery had all the latest design features: running its length were three rectangular skylights with chamfered corners, roof trusses were hidden and the walls were uninterrupted by pillars or windows.[23] At the far end of the gallery, at right angles to its central line, was a storage area for pictures.[24] One wall only was broken by a central fireplace. Some of these details were prefigured in sketches made by Turner in the *Tabley No. 3* sketchbook (in use in 1808; fig. 21), in which the artist also notes a thrifty heating method by reusing heat that would otherwise be wasted: 'Flues from the Back parlour or Kitchen warm the Gallery. Ventilation of the Gallery – and the Blinds to bow and set in behind a moulding to exclude the sun's rays.'

Turner's competence and self-confidence as an architect was total, and had circumstances been different, he would have made a creditable Professor of Architecture at the Academy. His vocation as a teacher, however, moved another way. He taught drawing and painting from the live model, and for years gave informal but much-respected advice on Varnishing Days. His

fig. 22 *Admiralty Screen and Building: Front View c.*1810 (perspective diagram: lecture illustration) (cat. 14)

greatest contribution was in the series of lectures as Professor of Perspective, which he gave from 1811 intermittently until 1828. In these he ploughed back to his students his knowledge of architecture and perspective, making some two hundred clear, large-scale drawings of perspective models to illuminate his often inaudible words and entrance his audience (figs. 22–4).

During the few years that Turner was planning Sandycombe Lodge, he was also devising his first set of six perspective lectures. It can be no coincidence that during this period architecture became a particularly vivid theme for him, spilling out into his work as an exhibiting painter from the practicalities of design and his own perspective research. *London* (exh. 1809; Tate Gallery), the pair of paintings of Tabley House, *Windy Day* and *Calm Morning* (both exh. 1809; University of Manchester and Petworth House respectively) and *High Street, Oxford* (exh. 1810; Private Collection) are all superb exercises in aerial and architectural perspective, and give form to his remark that 'Perspective may justly be

considered the colouring of Architecture.'[25] For his own reasons, he delayed beginning his first lecture series until January 1811, just over three years after his appointment as Professor of Perspective, and in exhibiting these pictures in the interim Turner was silently displaying his credentials as the appointed but as-yet-untested professor.

To plan his lectures and the diagrams that went with them, Turner walked around London making small perspective sketches of buildings.[26] He read widely and wrote up the first drafts of his lectures. His initial notes went into what has become known as his *Perspective* sketchbook (fig. 25), a little red morocco pocket book full of diagrams and intense notes on perspective, the action of light, of colour, reflection and art history, interspersed with pages of poetry.

Although Turner's work as Professor of Perspective has been thoroughly examined elsewhere,[27] there are aspects of his engagement with the science of the subject that should be looked at here. The first point to make is that the published sources he used were all of the eighteenth century or earlier. He studied, and taught from, the Renaissance theorists Guidobaldo del Monte

fig. 23 *Interior of a Prison c.*1810 (perspective diagram: lecture illustration) (cat. 15)

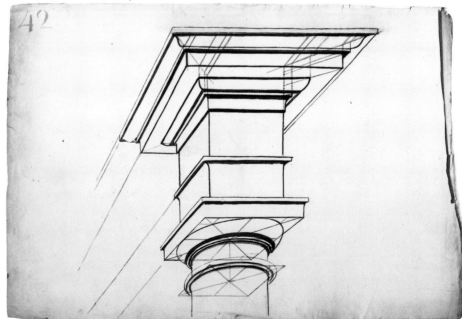

fig. 24 *Entablature of a Tuscan Column, with Perspective Lines c.*1810 (perspective diagram: lecture illustration) (cat. 16)

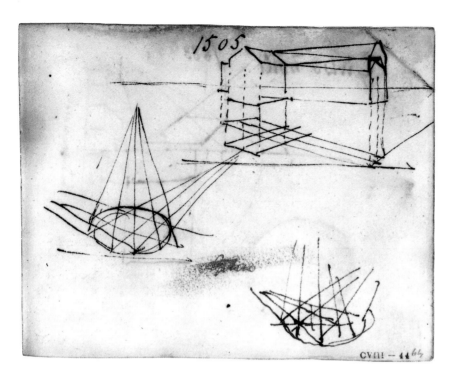

fig. 25 *Notes on Architectural Perspective* 1809, from *Perspective* sketchbook (cat. 20)

(*Perspectivae*, 1600), Salomon de Caus (*La Perspective avec la raison des ombres et mirroirs*, 1612), Giovanni Paolo Lomazzo (*A Tracte Containing the Artes of Curious Paintinge, Carvinge and Buildinge*, trans. R. Haydocke, 1598) and Lorenzo Siringatti (*Le pratica della prospettiva*, 1596). He used Samuel Cunn's 1759 edition of *Euclid's Elements of Geometry*[28] and also made extensive references and diagrams from seventeenth- and eighteenth-century English perspective theorists including John Hamilton, Joseph Highmore, Joshua Kirby, Thomas Malton, Joseph Moxon, Joseph Priestley and Brook Taylor. In looking back at the theories of the generation of his own teachers and earlier, while developing his own ideas for his students, Turner is positioning himself as the leading perspective theorist of his day. His technique was to make resumés of past theories, and to compare and contrast different approaches. In his clear understanding of the subject, Turner shows himself to be a skilled geometer, as comfortable in expressing the static theorems of Euclid, as he was with painting the motions of the wind and the sea.

Turner's engagement with architecture was interwoven by close friendships with half a dozen or more architects. Central among these was Sir John Soane (1753–1837), a good friend for the whole of Turner's professional life until Soane's death. What we know of Turner's own architectural manner was greatly indebted to Soane in its spare, elegant classic detail, with rounded arches and strapwork. Another architect friend was Jeffry Wyatt (1766–1840), a nephew of James Wyatt, who from 1800 to 1803 lived next door to Turner at 65 Harley Street[29] and later had a temporary home in the Winchester Tower of Windsor Castle. The friendship with Wyatt, who changed his name to Wyatville in 1824, was longstanding and extended to Turner visiting him annually at Windsor during the later 1820s, when Wyatville lived on site during the design and construction of St George's Hall.[30] The pair shared a roughness of dialect, both were short and stocky, and both lacked the courtier's egregious charm.

The most influential architect of Turner's generation, C. R. Cockerell (1788–1863), engaged the artist in 1821 on a plan to make a series of fifteen watercolour drawings of the Temple of Jupiter Panellenius at Aegina. This was an informal agreement, and although it never came anywhere near completion, it did generate long and fruitful conversations between the two men. Cockerell kept an extensive diary which throws light on their compatibility: 'Turner came to look at my drawings. Did not find him so methodical or stiff as he seemed before. Liked the subjects. Stayed from 10 til 9. It was a vast pleasure for me to look over my views with a man who felt them as he did.'[31] Four years later, though their joint project had not moved on, Cockerell found Turner to be just as loquacious and informed as before: 'visited Turner; found he had done little or nothing ... stood more than two hours with him talking of Vanburgh, Hawksmoor, & other he as usual standing with his hat on.'[32]

Other architect friends included Thomas Allason (1790–1852), Thomas L. Donaldson (1795–1885), James Hakewill (1778–1843), Philip Hardwick (1792–1870), the son of Thomas Hardwick the Younger, and John Nash (1752–1835). Allason, Donaldson and Philip Hardwick had all been Turner's pupils at the Academy, and no doubt listened carefully to what he had to say, for the quality of their perspective drawings was of the highest order. Allason, for example, invited Turner to collaborate with him on the set of published engravings *Picturesque Views of the Antiquities of Pola, in Istria* (1819),[33] while Turner and Hakewill worked together on the illustrations to *A Picturesque Tour of Italy* (1818–20).[34]

During the period of their collaboration on *A Picturesque Tour* Hakewill gave Turner information about travel in Italy, and wrote advice on getting about in the country into one of Turner's notebooks.[35] The relationships with some of these men extended into other areas: Turner met Donaldson again in Rome in 1819 and travelled with him to Naples; he engaged Allason to survey and value his inherited properties in Wapping, and drew him in when he had legal difficulties with his tenants;[36] Nash was a good friend and host to Turner; and Philip Hardwick remained close all of Turner's life, ultimately becoming one of the artist's executors.

A mutual friend to many of this circle was the water-colour painter and collector James Holworthy (1781–1841), who in 1825 and 1826 was overseeing the building of his own house at Hathersage, north Derbyshire. To him Turner expressed some of his most revealing thoughts about his attitude to, and delight in, the process of building:

fig. 26 *Landscape with Workmen Building a Bridge c.*1817, from *Durham, North Shore* sketchbook (cat. 22)

fig. 27 *Waterloo Bridge from Beneath c.*1818–20, from *Scotland and London* sketchbook (cat. 23)

you talk of mountains high as the moon, and the creaking timber wain labouring up the steep … but consider the pleasure of being your own architect day by day, its growing honors hour by hour, increasing strata by strata, but not the clang of the trowel I fear, minute by minute, according to your account.[37]

This freely expressed joy in the act of construction is expressed repeatedly in Turner's paintings and sketches. A year or two after he had completed Sandycombe Lodge, he painted *Dido Building Carthage* (exh. 1815; National Gallery, London). Quite apart from the fact that the whole conception of the painting from its title outwards concerns building, and develops the metaphor of building a nation, there is in the bottom left-hand corner of *Dido* a detail of a pile of bricks of the kind that Turner will have seen every time he paid a site visit to

Sandycombe. The sketchbooks abound with insights into the building industry of the early nineteenth century (figs. 26, 27). As in Turner's boyhood, so in the years after Waterloo London came alive again to new construction. This was noticed in 1822 by J. Davies, an architect just setting up in London after some years on the Continent:

London in some places I find very much changed. Not the City Cheapside the Strand &c are the same scene of bustle and confusion as ever but the *West End* is a new world to me. Canals have been cut. Where used to be green fields I now see the masts of vessels. New roads new streets new crescents, terraces, squares, churches &c &c the change is wonderful. London is rendered more vast more magnificent but still there is rather more of bricklaying than of Architecture.[38]

The most technologically advanced, and visually exciting, kind of construction in London during this period, were the new bridges over the Thames. Vauxhall, Waterloo and Southwark Bridges were all

fig. 28 *Study of Fortifications at Laurenstein* 1817, from *Waterloo and Rhine* sketchbook (cat. 21)

fig. 29 *Study for 'Rome from the Vatican'* 1819 (cat. 17)

built between 1813 and 1819, when it became an economic imperative to improve access to the south of London, with its growing population, its new industries and its proximity to the naval and civil ports on the south coast. Hammersmith Bridge, the first suspension bridge in the capital, was built to the designs of William Tierney Clark and opened in 1827. The renewed London Bridge was opened in 1831 (see below, Chapter 6). When construction on this began, Thomas Donaldson described the scene in a letter to Robert Finch:

They have already begun upon London Bridge which is to consist of 5 eliptic arches somewhat as here drawn [sketch]. They are driving Piles. Scientific men profitting by the excess of money in the market are devising schemes which tend at the same time to public utility & individual profit & by this means we have some very useful projects in contemplation. There is one of a suspension Bridge near the Tower which will allow vessels of 200 tons to pass up the River without striking their masts. This sketch will give you an idea of what it is to be: it will be Iron of course and about ⅝ of a mile in length each arch 600 ft span.[39]

The suspension bridge near the Tower was never built, but the burgeoning of public works in London after Waterloo, directed towards the expansion of trade around the globe, was vivid evidence of London becoming the capital of the world. Drawn as he was to the banks of the Thames, Turner made notes of its bridges, particularly when they were under construction (figs. 27, 107–8), and of the bustle of the river around them. As evidence of man's good management of the landscape, and as significant compositional devices in themselves, bridges continued to be an important feature of Turner's landscape subjects. Not only were they obligatory devices in his Claudean or Poussinesque subjects, such as *The Opening of the Vintage of Macon* (exh. 1803; Sheffield Art Galleries) or *Apullia in Search of Appulus* (exh. 1814; Tate Gallery), but he gave them added metaphorical significance in *Crossing the Brook* (exh. 1815), *Kirkstall Lock on the River Aire* (1824–5) and *Rain, Steam, and Speed – the Great Western Railway* (exh. 1844) (figs. 92, 98, 104). In other drawings, such as the group of studies around Freiburg (1841; fig. 110), the suspension bridge is the natural focus of the compositions. Using the bridge as a marker, Turner creates a strong sense of place and modernity in the drawings, and makes a dynamic tension between the fragile bridge and the surrounding town and mountains which is quite breathtaking.

Other large-scale public works caught Turner's eye on his European tours. The harbour of Honfleur[40] seen from above is a stoutly built, tight little township protected by massive quays and sea defences. Turner went even further, perhaps risking arrest as a spy, in making careful drawings of the fortifications at Laurenstein in the *Waterloo and Rhine* sketchbook of 1817 (fig. 28), annotated with the names of architectural components – glacis, counterscarp, ditch, rampart and banquette. His studies of Rome, made in 1819, include *Rome, from the Vatican* (fig. 29), a working document in which Turner recorded the architectural information he needed for the grand painting he was planning for the 1820 Academy exhibition, which came to be called *Rome, from the Vatican. Raffaelle, Accompanied by La Fornarina, Preparing his Pictures for the Decoration of the Loggia* (Tate Gallery). The determination and firmness of the drawing sets it apart from other evocative light-filled studies that Turner made on this journey, and reveals it as the tool that it is, to record the artist's full understanding of the complexities and uses of architecture, as direct reference for the painting to come.

Such passion for building blocks, pillars and beams lasted until the end of Turner's life. In two of his last surviving letters he speaks of watching the construction of the Crystal Palace for the Great Exhibition. Writing in December 1850 to Hawksworth Fawkes he observed:

The Crystal Palace is proceeding slowly I think considering the time, but suppose the Glass work is partially in store, for the vast Conservatory all looks confusion worse confounded. The Commissioners are now busy in minor details of stowage and hutting, all sent before the Glass Conservatory is ready – to be in bond under the duty to be lay'd on if sold.[41]

A month later he wrote to Hawksworth again:

The Crystal Palace has assumed its wonted shape and size. It is situated close to the Barracks at Knights Bridge, between the two roads at Kensington and not far from the Serpentine: it looks very well in front because the transept takes a centre like a dome, but sideways ribs of Glass frame work only Towering over the Galleries like a Giant.[42]

Here are words which may stand as Turner's epitaph: 'Towering over the Galleries like a Giant.'

3
THE OLD TECHNOLOGY: EARLY EXPERIENCE OF INDUSTRY 1775–1815

Turner had a lucky boyhood. By accident of birth he grew up in a part of London where industry and commerce, and the worlds of art and the theatre, shipping and mechanical engineering, met within an area of half a square mile. All this was in the same geographical orbit as the learned societies discussed in Chapter 1, and in addition to it. Covent Garden was the city's main fruit, flower and vegetable market, humming with activity around the clock. It was also the home of two of London's great theatres, whose performances brought thousands to the area nightly. Long Acre, which marked the north-western edge of Covent Garden, was the centre of Britain's road transport industry. Here, in back lots, ground-floor workshops and out on the street carriages and conveyances of all kinds were designed and built. There were iron foundries, carpentry shops, upholsterers, sign painters and colour makers in Long Acre, all mixed up with the other suppliers and manufactures that make a town spin.[1] A few hundred yards south lay the river, busy with traffic passing by and loading and unloading at the wharfs and jetties, and at the new stone quay at the Adelphi. In Brentford in the mid-1780s, the butcher's shop owned by Turner's uncle backed directly onto the River Brent, a tributary of the Thames, and there, too, Turner will have seen constant comings and goings of trading craft from the villages north-west of London and further afield. In 1791, as a sixteen-year-old apprentice, Turner spent some months painting stage sets in the Pantheon Opera House, where he would have seen the stage machinery static and in action.[2]

These impressions came perhaps too early to make it into the sketchbooks. In the 1790s, however, when Turner began the life-long series of journeys around Britain that were to provide the staple for his inspiration, his sketchbooks and finished drawings reveal the rate at which the nature of his interests as an artist developed. The first journey to be extensively documented, to Bristol in 1791, resulted in a collection of views of the Avon Gorge, St Mary Redcliffe and country houses and abbeys near Bristol.[3] These were essentially topographical, of the kind which fed the market for engraved views, and had no particular industrial edge to them. During the next few years, however, the pace quickened, and on journeys to Dover, the Isle of Wight, Wales, Hereford and Derbyshire Turner's eye began to be caught by industrial subject matter.

The process of widening perception seems to have begun with the sixteen- or seventeen-year-old Turner being gradually struck by the *curiosity* of much of what he saw. In Dover around 1793, for example, he made a series of pencil drawings of the harbour landing stage, with dockside paraphernalia such as windlasses, flagpoles and so on. There is one extraordinary drawing from this group of the masts of sailing ships towering up from behind some quayside houses,[4] and another of a heavy sea rising up dangerously between a sailing boat making for harbour and the fragile pier.[5] During his first tour of the Midlands in 1794 and to Hampshire and the Isle of Wight in 1795 Turner's eye began to mature. He seems now to pick out watermills and windmills, foundries and forges, identifying them as the focuses of what it is that makes the landscape work. This was not at the expense of other subject matter, but industrial technology did seep into his sketchbook studies at this time, if not so much into the subjects he worked up for sale as watercolours, and there must be reasons.

One impulse may have been Turner's contact in 1792 and 1793 with the Society for the Encouragement of Arts. Among the inventions and improvements submitted in the year that Turner won his Silver Palette were models of wharf cranes, bolt-driving machines for use in shipbuilding and machines for raising ore from mines.[6] The Society's house in John Adam Street, where Turner sat his drawing examination in April 1793, was a centre of discussion and display of technological improvements. Here, practical technology met theoretical science, as inventors and manufacturers came together to air new ideas and study the latest drawings and models.

At the same time as he began to draw industrial subjects, Turner also developed an interest in depicting people at work that was to last all his life. This started early: in one of the watercolours made in Bristol in 1791 or very soon after, *The Rising Squall: Hot Wells from St Vincent's Rock, Bristol* (Bristol City Museum and Art Gallery), sailors are shouting and trying quickly to tie up a boat while another boat runs at speed towards them before the wind. Work here is intertwined with drama.

In *The Pantheon, the Morning after the Fire* (exh. 1792) the topical narrative is driven by urgent work and gestural action.[7] Sketchbook studies show something of the roots of these interests – men ploughing, pulling on ropes, standing around at fairs, struggling with their boats – everyday life in the late eighteenth century which Turner took careful note of and tucked away.[8]

It is a small step, and all part of the same integrated thought process, from drawing people at work to drawing their places of work. In early Turner there are two parallel channels that reveal how his interest in indus-

fig. 30 *Marford Mill, Denbighshire* 1794 (cat. 26)

trial subjects evolved. The first is his depiction of mills as Picturesque subjects; the second is his attitude as an artist of the Sublime to the drama of foundries and forges. By the late eighteenth century windmills and watermills were the old technology, being gradually superseded by steam engines. Rural watermills, particularly those in or near the iron-producing areas of south Wales, Shropshire and Yorkshire, were more vulnerable to change than windmills. When they were abandoned as uneconomic, watermills fell to pieces and were left to rot. Turner's depictions of them reveal what is perhaps the beginnings of this change. Although some, such as (?)*Honiton Mill*[9] and *Pembury Mill, Kent* (c.1795–6)[10] show vigorous industrial activity and relative prosperity, others are drawn as if they were already ancient monu-

ments. *Marford Mill, Denbighshire* (fig. 30), has one wheel tumbled into the undergrowth and is of a distinctly erratic construction. In his earliest known painting in oil, *Watermill and Stream* (c.1791–2; fig. 31), Turner paints the subject as a decorative image, with no sense of place, oval-shaped in the manner of Gilpin's illustrations to his treatises on Picturesque travel. Whether prosperous or down at heel, descriptive or decorative, Turner's watermills are always aged, venerable structures.

Windmills, however, necessarily on higher ground and in arable areas, continued throughout the nineteenth century as powerful symbols of civilisation, order and economic prosperity. In Turner's drawings of the late 1790s they characteristically appear proud, dominant and in good repair. One windmill (fig. 32)[11] stands above a group of well-ordered cottages with clouds scudding by, while in *Nunwell and Brading from Bembridge Mill* (1795; fig. 33) Turner gives a very clear account of how the mill works, and shows it in the process of being re-sailed. The difference in Turner's approaches to watermills and windmills is a measure of the reason he brought to his choice and handling of industrial subjects.

With drawings of foundries and forges Turner places yet different emphases. From the inside these are low-toned, dark places with sudden lights of sun or fire. The noises that come from them are hard and intermittent, the clang of a trip hammer every ten seconds or so superimposed on the constant rush of the watermill that drives the machinery.[12] In *An Iron Foundry* (*c*.1797; fig. 34) the centre of attention is naturally the great tilt hammer and the red hot iron. Turner paints things as they are – the gnarled old beams, the leaning frame of the building, the water rushing in at one side and out at the other – and in doing so he suggests not only the scale of the operation in relation to the size of the men working it, but also the overwhelming noise they had to contend with. His immediate forebears in subject matter of this kind are Joseph Wright of Derby (1734–97) and Philippe Jacques de Loutherbourg (1740–1812), and while he

fig. 32 *Cottages and a Windmill (location unknown)* mid 1790s pencil and watercolour (TB XXVII P)

fig. 33 *Nunwell and Brading from Bembridge Mill* 1795, from *Isle of Wight* sketchbook (cat. 40)

fig. 34 *An Iron Foundry* 1798 (cat. 30)

shares with them an excitement at the dramatic lighting effects, Turner alone conveys the sense of the dirtiness of the industry. New though the process may be, there is also in Turner an evocative sense of timelessness, of its having been carried on beyond memory.

Joseph Wright, and to a lesser extent de Loutherbourg, used the language of the old masters to convey a sense of mystery and antiquity in their industrial subjects. The roots of Wright's *An Iron Forge* (1772; Tate Gallery), for example, lie both in Rembrandt and in quattrocento Italian art. For Turner, Rembrandt was one of the select group of the greatest old masters against whom he measured himself throughout his life.[13] Among the paintings of Rembrandt which Turner knew well was *Landscape with the Rest on the Flight to Egypt* (1647; fig. 35), which he will have seen in the collection of Sir Richard Colt Hoare at Stourhead when he went there to carry out a commission in 1795. *Limekiln at Coalbrookdale*

(fig. 36) is a direct transcription of *Rest on the Flight* into industrial terms. There are some obvious parallels – the proportions, the composition, the two sources of light, the strong chiaroscuro, the air of timeless mystery.

There is, however, enough in Turner's painting to show that, while taking Rembrandt's influence, he is shaking off the religious connotations that so thoroughly infused the approach of Wright's generation, and allowing the subject to speak for itself. Men had been burning lime for centuries – what they did (and still do) was to roast chalk or limestone, calcium carbonate, to drive off carbon dioxide and leave calcium oxide, or lime. This was used extensively in building and as a fertiliser, and is one of the oldest chemical reactions known and practised by man. Lime-burning is patently not a new process of the Industrial Revolution. The moderately large furnace shown here is of the kind that had been in use for generations, and in going to

fig. 35 Rembrandt *Landscape with the Rest on the Flight into Egypt* 1647 oil on panel *National Gallery of Ireland*

fig. 36 *Limekiln at Coalbrookdale* c.1797 (cat. 27)

Coalbrookdale and returning to paint not the new iron-works but an ancient process, Turner raises some serious questions.

The painting, which was never exhibited at the Royal Academy, had no known title until 1825 when it was engraved by F.C. Lewis as *Colebrook Dale*, evidently with Turner's approval.[14] There are, however, no identifying features in the landscape, which will have been known to Turner not only through his visits but also through the group of small pen studies of industrial sites in Shropshire and south Wales by de Loutherbourg (fig. 37). These, in the 1790s, were in Dr Thomas Monro's collection, and in 1833 Turner bought some, if not all of them, at auction.

Whether the source is Rembrandt, de Loutherbourg, his own observations or a wholesale amalgam, Turner's intention in *Limekiln at Coalbrookdale* appears to be to use an old master's manner to depict an ancient practice,

and thus underline the fact, to a generation who would know these things, that despite the advances of the Industrial Revolution lime-burning has been going on for centuries. This does not deny Turner's clear-eyed interest in modern industrial practices, but in choosing to enshrine his *Limekiln* in oil on panel rather than the more ephemeral medium of watercolour, and to create it with a Rembrandtian voice, he is defining the eternity, as opposed to the modernity, of industry.

During his 1798 tour to Wales Turner drew the series of four large-scale, partially completed, pencil views of the iron-working community of Cyfarthfa, near Merthyr Tydfil (figs. 38, 39). In the late 1790s Cyfarthfa was owned by Richard Crawshay (1739–1810), an iron manufacturer and merchant who traded both from Cyfarthfa and from George Yard in the City of London. The ironworks had been founded in the mid-1760s by Anthony Bacon (1718–86), Member of Parliament for Aylesbury, and William Brownrigg, and grew to become one of the largest suppliers of cannon and other

fig. 37
Philippe Jacques de
Loutherbourg,
*Welsh & Shropshire
Industrial Subjects:
'Fire Engin Coalbrook Dale'*
(cat. 39)

fig. 38 *Cyfarthfa Iron Works* 1798 (cat. 28)

ordnance to the army and navy in the early years of the Napoleonic Wars.[15] Brownrigg was replaced as Bacon's partner by Crawshay in 1777, and on Bacon's death in 1786 his share of the Cyfarthfa business passed to his illegitimate son Anthony Bacon the younger (1772–1827).[16] This Anthony Bacon sold his interest in Cyfarthfa to Richard Crawshay in 1794, though he retained a royalty, and later surviving correspondence shows that he kept an interest in the running of the company at least in the 1810s.[17]

The drawings were commissioned by the younger Anthony Bacon, who lived near Newbury, and at his death retained property in Glamorganshire.[18] There is a specific order, written in Bacon's hand on the back flyleaf of the 'Hereford Court' sketchbook:

Anthony Bacon near Newbury Berks
4 drawings of the Iron Works of Richd Crawshay Esq at Cyfarthfa near Merthyr Tidvil – 18 miles from Cardiff – 16 from Brecon – 10 [9½ written above] by 13 inches – 5 gns Each.[19]

As this is Bacon's hand it reveals the contemporary custom of making a binding contract, the commissioner confirming it by writing the full agreed details into the artist's notebook. It was Turner's practice at this time to make sketches followed by detailed drawings on the spot. These would be carefully coloured back in his studio in London, and comprise the commissioned work. Figs. 38 and 39 are detailed drawings, not initial sketches, and that they have no trace of colour strongly suggests that the commission was called off during its course for reasons we do not know.

The state that the drawings were left in throws a little more light on Turner's practice in the field. The main industrial and topographical interest in them is in the central band running across each drawing, and, in each case, the foregrounds and backgrounds are unworked. This parallels Turner's method of taking notes of architecture on his travels (see Chapter 2 above), in which he would draw enough of the detail of a building to allow him to *ad lib* at home with the rest. In the *Cyfarthfa* series

fig. 39 *Cyfarthfa Iron Works* 1798 (cat. 29)

fig. 40 *Forge Scene* 1796–7, from *Wilson* sketchbook (cat. 41)

he left himself with the opportunity to fill in the fore-grounds later as the fancy – or his stock of sketchbook details – took him. We can take this thought a little further, and suggest that there were times when Turner purposefully left these spaces blank for the later growth of the incidental detail that fills the foregrounds of many of the *Picturesque Views* watercolours. Thus, they grew out of Turner's intellect as much as his observation – the product of ratiocination as he painted quietly at home – and were not necessarily related to what he saw on the day he sat in front of the subject with his pencil and paper.[20]

The *Cyfarthfa* drawings are within the tradition of the late eighteenth-century panoramic landscape, as expressed by Paul Sandby, Thomas Hearne and Thomas Girtin. Turner, however, seems to be extending the convention by revealing how the complicated pattern of buildings fitted together, and how they in turn knitted themselves around the industrial paraphernalia of Cyfarthfa, its big waterwheel, chimneys, kilns and elevated waterway. Although we may not be shown the

inside of the buildings, we can see that Turner's *Cyfarthfa* is accurate, and we can almost attempt a ground plan. Comparison with Richard Pamplin's views of the iron-works in the 1790s (Cyfarthfa Castle Museum and Art Gallery, Merthyr Tydfil) confirms the veracity of Turner's topographical record.

Two sketchbooks in use in the late 1790s, the *Wilson* and *Swans* (figs. 40, 41), reveal how omnivorous Turner's interests as an artist were at this time. TB XXXVII was called the 'Wilson' by Ruskin when he first catalogued the Turner Bequest because it bore Turner's label 'Studies for Pictures. Copies of Wilson'. The label is not a tautology, as Ruskin's too brief, and misleading, title might imply. 'Studies for Pictures' refers to compositions which Turner himself was contemplating, while 'Copies of Wilson' speaks of Turner's own transcriptions in the sketchbook of some paintings by Richard Wilson. The compositions in the 'Wilson' sketchbook are generally worked up to a much greater degree than in any other sketchbook in the Turner Bequest. There is a lighthouse with the sun setting over the sea;[21] a view of the bay at Margate, followed by four double-page spreads of the interior of a church;[22] a view of a harbour – possibly

fig. 41 *Foundry with Tilt Hammer* c.1798, from *Swans* sketchbook (cat. 42)

Margate – seen from a boat;[23] and a scene inside a forge with anchors lying about (fig. 40). All could be quite comfortably scaled up to exhibition size, either in oil or watercolour, and the variety of type of study reveals how industrial subjects settled naturally but unobtrusively within the wide range of his interests. As Jack Lindsay succinctly put it in 1966, 'despite his extreme interest in men at work and their tools, Turner did not deal with the factory system directly to any extent. Still, he did not ignore it.'[24]

In the *Swans* sketchbook, one of the books which Turner took to Bristol and South Wales on his 1798 trip, there are some evocative pencil studies of Bristol[25] which compare with the studies of Birmingham and Coventry made more than thirty years later (fig. 102, cat.98). Some pages further on there is a luminous, well worked-up pen-and-ink drawing, extending over a double page, *Foundry with Tilt Hammer* (fig. 41), which is probably related to *An Iron Foundry* (fig. 34).

From about 1800 until around 1806 industrial subject matter seems to fall away in Turner's oeuvre. He was concerned in these years with becoming a full member of the Royal Academy; with wooing patrons, and travelling to Scotland (1801) and Switzerland (1802); with creating a pictorial language of classicism which would vie with Claude and Poussin; and with designing and building his own picture gallery where he could show his work to its best advantage. Turner had a huge range of other interests as a painter – his marine subjects were attracting some of the highest people in the land, and in

fig. 42 *A Country Blacksmith Disputing upon the Price of Iron, and the Price Charged to the Butcher for Shoeing his Poney* exh. 1807 (cat. 32)

watercolour he was searching for new interpretations of the wildness and variety of nature.

During these five or six years when Turner was establishing himself as the archetypal professional painter, industrial subjects were no longer fashionable with collectors. When they caught his eye, they remained in his sketchbook. Thus, during his journey to Scotland in 1801, he spotted a blacksmith's shop at 'Gretny Green' [*sic*];[26] he noted local people in their curious tartan clothes, but they were only of passing interest.[27] It was not until around 1804 and 1805, when he had fully expressed the excitement of his continental tour in a dozen or more large canvases and many more watercolours, that Turner began to find time to relax and ruminate in the landscape of the Thames valley. There

he drew details of watermills and people working machinery;[28] interiors of a blacksmith's shop;[29] and people getting in the harvest.[30] These are all quick, passing studies, but nevertheless they are there. Inscriptions against some of them show that Turner was again carefully watching what people do: 'breaking off sticks and putting them in the grate … girl filling the tea kettle out of a large Brown jug';[31] 'Bargeman hanging up cloats &c on the shrouds … good moment [?] to avoid the long line of shrouds.'[32]

Information of this kind was building up in his memory when an external event came to suggest that he might now do something specific with it. The young Scots painter David Wilkie (1785–1841) had reintroduced the fashion for low-life painting at the 1806 Academy exhibition. His painting *The Village Politicians* caused the kind of stir that Turner had come to believe was reserved for him. It has been argued that this public acclaim for an artist younger than he caused Turner to choose a new kind of subject for exhibition. Thus, the following year, he exhibited *A Country Blacksmith Disputing upon the Price of Iron, and the Price Charged to the Butcher for Shoeing his Poney* (fig. 42), a subject quite at odds with anything else he had shown in the past. There were current political issues which may have effected Turner's choice of subject in *A Country Blacksmith*. The government had in May 1806 set a new tax of £1 a ton on all iron products made in Great Britain, and this gradually had a direct effect on local prices, right down to the increase in cost of the nails in a pony's shoe.[33]

However it came about, the fact is that Turner looked seriously once again at industrial subject matter during the period 1806 until about 1810. While picking up incidental details of everyday life, the sketchbooks of around 1804–7 and other informal studies reveal Turner to be formulating compositional ideas every bit as serious as the non-industrial compositions of the same period. Thus, among the studies for the *Liber Studiorum* there is a *Blacksmith's Shop* (fig. 43) of Rembrandtian darks and lights, and the *Harvest Home* and *River* sketchbooks contain studies made on the estate of Lord Essex at Cassiobury Park of figures for the oil paintings *Harvest Home* and *Reaping*.[34]

It is likely that at this time Turner painted the interior of a forge which about forty years later he reworked as *The Hero of a Hundred Fights* (fig. 115). From what little we can now see of its original state – for Turner obliterated nearly half of its surface in 1847 – it has all the ingredients of a considered painting for exhibition at the Academy: imposing machinery, moodily lit from the side, a woman seated with a basket, and a litter of

fig. 43 *Blacksmith's Shop* c.1806–10 (cat. 31)

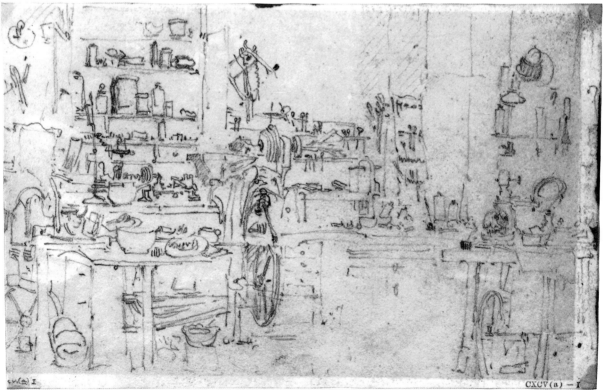

fig. 46 *Composition Study (with Tripod and Pulley) for 'Dorchester Mead, Oxfordshire'* 1805–9, from *Hesperides II* sketchbook (cat. 44)

utensils, fruit and vegetables. Why was the picture never exhibited? Perhaps, like the *Cyfarthfa* series, it was a commission that failed; perhaps, indeed, it was a delayed response or extension to the *Cyfarthfa* commission. A further explanation lies in the fact that Turner was the butt of some intense private criticism of his low-life pictures at this time, and withdrew them. Thomas Hearne, with Farington and Edridge, visited Cassiobury Park in July 1809 and scorned the 200 guineas that Lord Essex had given for Turner's *Trout Fishing in the Dee* (exh. 1809; Taft Museum, Cincinnati). Discussing what might be a fair price for a Turner, Hearne said: 'For a person a full admirer of his pictures 50 guineas, but for myself I would not give fifteen.'[35] This may have been only one of a number of direct or indirect critical attacks, to which Turner was particularly sensitive. Of *The Unpaid Bill*,

another low-life subject exhibited in 1808, a critic wrote that the figures were 'wretchedly drawn', and that 'for a picture of colouring and effect, it is not only unexceptionable but inestimable'.[36] Many years later, when Lady Eastlake had visited Turner's Gallery, she saw *Harvest Home* and reported that Turner had been 'disgusted at some remarks' about it, and never finished it.[37]

Another abandoned low-life subject of the period is the curious painting *An Artist's Colourman's Workshop* (fig. 44). This, like the unnamed and unexhibited forge interior (fig. 115), has shadowy mechanical devices with crepuscular activity in the foreground. There is no clue to where the idea might have come from – apart from the fact that the subject must have been prompted by any one of the number of colourmen's workshops which supplied Turner with paints[38] – but with its clutter of tools of the trade, emerging anecdote and other incidental detail the painting shares something of the character

(*left, top*) fig. 44 *An Artists' Colourman's Workshop* c.1807 (cat. 33)

(*left, bottom*) fig. 45 *Interior of a Workshop* 1807–8, from *River and Margate* sketchbook (cat. 43)

fig. 47 *Dorchester Mead, Oxfordshire* exh. 1810 (cat. 34)

of *A Country Blacksmith* and *The Unpaid Bill*. In the *Chemistry and 'Apuleia'* sketchbook,[39] in use in 1813, there are sixteen pages of closely written pigment and varnish recipes which suggest that by this time Turner was attempting to make his own materials, and withdrawing to some extent from the services and expense of colourmen.

The narrative drive in these so-called low-life paintings evolved away from anecdote, towards a quality of poetry and universal meaning that takes its place naturally within the classical subjects of Turner's early maturity. What may have begun with a bang, did not end with a whimper but with the natural confluence of the tributary of low-life anecdotal subjects into the mainstream of Turner's classicism. *Dorchester Mead, Oxfordshire* (exh. 1810; fig. 47) returns to the open air, takes the low-life manner to pieces, and reconstructs it in a new pictorial language. But while painting in terms of Cuyp and Claude – the grey mist, the rhythm of arches, the classical structure, the quiet cows – Turner has created a thoroughly contemporary workaday scene. Logs are being unloaded from a cart with a tripod and pulley for onward transport by barges, and a hay cart trundles along the high road beyond. Behind the poetry lies the reality of working life – if the logs are not loaded by nightfall, the barge cannot sail at dawn, and somebody does not get paid. In these small vignettes within a larger subject, Turner sticks the splinter under the nail, and shows that the landscape has to be worked for its inhabitants to stay alive.

In the weeks after *Dorchester Mead, Oxfordshire* was exhibited, Turner travelled to Rosehill in Sussex to stay with John Fuller MP (fig. 3) and to carry out a commission to make drawings for an oil painting of his house.[40] Fuller owned sugar plantations in Jamaica, and these brought him immense wealth. At home, however, he had inherited in 1777 Rosehill, incorporating nearly six thousand acres of Sussex farm and woodland which had provided the fuel for the family ironworks and wood for ship-building. Until the third quarter of the eighteenth century Sussex had been one of the foremost iron-producing regions in the country, making cannon for the navy in Portsmouth and Chatham, and domestic ironware for London and the south of England. By the time John Fuller inherited Rosehill, however, the family forges had long failed to keep up with new industrial techniques and worked irregularly, the victim of their own mismanagement and of the innovations at Cyfarthfa, Coalbrookdale and elsewhere. In 1787

Fuller's Heathfield forge closed, leaving, as a result of over-felling, wide tracts of bare downland.[41] The neighbouring Ashburnham forge struggled on until 1813.[42] Nonetheless, Fuller had enough surplus income to cushion him from the immediate effects of the industrial decline, and he spent heavily on embellishing Rosehill. He commissioned Turner over an extended period, and invited Sir Robert Smirke to design an observatory on Brightling Hill equipped with new astronomical instruments. Fuller lived his life on a grand scale, being not only a patron of the arts and a generous host, but also a staunch supporter of the Royal Institution from its foundation (see Chapter 4 below), and its benefactor at his death.

The Vale of Ashburnham (1816; fig. 48) is one of twelve watercolour subjects commissioned from Turner by Fuller.[43] These were intended at the outset to be engraved for wider circulation, like *The Southern Coast* series, and in them lie the beginnings of the long series

fig. 48 *Vale of Ashburnham* 1816 (cat. 36)

of *Picturesque Views of England and Wales* that became Turner's interior odyssey.[44] The landscape becomes an integrated whole from the foreground to the distant Beachy Head, the curve of the bay and of the hills and valleys being echoed in the forms of the foreground tree trunks. The subtext of the watercolour – trees and the shrinking of woodland – infuses every detail, and frames Ashburnham Place, the home of the Earl of Ashburnham, an only marginally more successful iron-founder than Fuller.

There is a lyricism in *Vale of Ashburnham* that deceives the viewer about the extent of the loss of woodland and about the underlying lament of the watercolour. Despite the fact that there is an ox cart in the foreground, apparently collecting wood to fuel the forge, Ashburnham forge was already closed. Another of the watercolours for Fuller, *Vale of Heathfield* (*c*.1815; British Museum – engraving, fig. 49), has a very similar composition, but in

place of the ox cart there are two rabbits playing. With these details, Turner may be underlining the fact that industry, recently vanished from Ashburnham, was long gone from Heathfield.

This decline is spelt out more clearly in the bleak drawing *Crowhurst* (*c*.1816; fig. 50) made for engraving in the *Liber Studiorum* series.[45] Though related to the Fuller watercolour *Pevensey Bay from Crowhurst Park*,[46] *Crowhurst* shows a landscape so much more ravaged and shorn of trees that it cannot solely be due to the predations of winter, the season depicted. The foreground figures, energetically sawing up some of the few remaining trees, seem driven to the destruction of their own livelihood. Turner did most of his travelling in the summer and autumn, so this winter subject is not only uncharacteristic but also, I suggest, a thoughtful commentary on the decline of the traditional industries of Sussex.

During his first tour of Devon and Cornwall in the

fig. 49 W.B. Cooke, after J.M.W. Turner, *Vale of Heathfield* 1818 (cat. 37)

fig. 50 *Crowhurst* c.1816 (cat. 38)

summer of 1811 Turner kept to the coastline, travelling in a clockwise direction around the peninsular. His purpose was to collect information for a new series of engravings commissioned by the engravers and publishers George and William Cooke. He observed landscape and the people in it, and how and where they lived. The uniting factors were of course the sea, which was always at hand, and Turner's emphasis throughout the series on geological formations, which will be looked at more fully in Chapter 7. The *North Devon* sketchbook has pencil studies of Tintagel Castle, with the extraordinary piece of machinery especially made to lower boats down to the sea there (fig. 51). In the years after his skirmish with low-life subjects Turner no longer associated machinery with darkness and mystery. Here, and in the watercolour painted in about 1815 (fig. 52) for

engraving (fig. 53), the mechanics of the capstan are set out clearly, so that its purpose is plain, to the extent indeed that it could be reconstructed from the evidence of Turner's drawing alone. A photograph taken c.1875 by Thorn of Bude shows how accurate was Turner's rendering of the capstan and its main purpose (fig. 54). This was to lower slate from the Delabole and Tintagel quarries down to ships waiting in Tintagel Haven below. Turner's sleight of hand, however, comes in his stretching of the scale of the cliff and the height of the drop, exaggerating his subject even from the earliest sketch (fig.51), while telling the absolute truth about it.[47]

(left) fig. 51
Tintagel Castle, with Lifting Gear 1811, from *North Devon* sketchbook (cat. 45)

(below) fig. 52
*Tintagel Castle c.*1815 watercolour. Gift of the Legatees under the will of Ellen T. Buliard in accordance with her request. Courtesy, Museum of Fine Arts, Boston

(right, top) fig. 53
Tintagel Castle 1818 (cat. 35)

(right, bottom) fig. 54
Loading slate at Tintagel *c.*1875. Photogaph by Thorn of Bude. *Royal Cornwall Museum, Truro*

4

OBSERVING THE SKY: METEOROLOGY, ASTRONOMY AND VISIONS

Towards the end of his life, Turner remarked:

When I was a boy I used to lie for hours on my back watching the skies, and then go home and paint them; and there was a stall in Soho Bazaar where they sold drawing materials, and they used to *buy* my skies. They gave me 1s 6d for the small ones and 3s 6d for the larger ones. There's many a young lady who's got my sky to her drawing.[1]

Turner is here referring to a period in the mid-1780s, a few years before he saw watercolours in quantity in the collection of Dr Thomas Monro, and so before he was able to amass much experience of how the older generation of artists, particularly Alexander and John Robert Cozens, approached the problem of painting skies.

Thus, Turner's lifelong observation of the formation and passage of clouds was rooted, like so much else in his work, in the experiences, enthusiasms and natural instincts of his childhood. His application to observation even then was heroic, and even if his boyhood habits are exaggerated in recall, this determination has echoes in written and drawn observation of the sky made by Turner in later life.

Studies that mark a turning point in the nature of Turner's observation of the sky are of a partial eclipse which he witnessed probably on 11 February 1804, when three-quarters of the sun's diameter was obscured in London (fig. 55). At further points of his life, these are echoed in diagrammatic pencil studies of the sun rising

fig. 55
Study of a Partial Eclipse of the Sun 1804, from *Eclipse* sketchbook (cat. 55)

fig. 56 *Schematic Memorandum of a Sunrise* 1822, from *King's Visit to Scotland* sketchbook (cat. 59)

pages, held with a brass clasp. The pages have all been given a tonal preparation of pale grey wash, which was already in place, of course, before Turner used it out on a hill somewhere to make the eclipse studies. Having made these studies across the first six openings of the book, he did not use it again. That this book was not reused is irrelevant, but the fact that he had it ready out in the country when the eclipse occurred suggests that he was waiting for the event and that he knew it would be an important thing to see. This is the earliest-known specific observation of an astronomical event in Turner, and clearly his curiosity about the coming phenomenon drove him to observe it. Eclipses were published well in advance in the *Nautical Almanac*, and Turner's apparent foreknowledge of this one suggests that his engagement with ships and the sea extended to his awareness of the value of sailor's tables.[3]

Henry Syer Trimmer, the eldest son of Turner's friend Henry Scott Trimmer, told Thornbury that as a child he, his father and Turner walked at Hammersmith

(1822; fig. 56) and setting (?1832; fig. 57) and setting again.[2] In February 1804 Turner was preparing for the opening of his private gallery in Harley Street. He was a Royal Academician with a high public reputation for works of genius and variety, and a private reputation amongst his colleagues as an unpredictable, if amiable man, with firm streaks of arrogance and bad temper. He was worldly, ambitious and had already revealed himself to be a great painter whose interests ranged widely about the observable world, and although an eclipse of the kind Turner drew in 1804 does not in the event appear in any finished paintings, it was there for recall in his sketchbooks.

The *Eclipse* sketchbook sets some interesting questions. It is an expensive, calf-bound book of one hundred

fig. 57 *Diagrammatic Progress of the Setting Sun* ?1832, from *Life Class* sketchbook (cat. 60)

fig. 58 *Eclipse, with a Gesticulating Figure* 1824 (cat. 48)

fig. 59 *Brightling Observatory* 1815–16, from *Hastings* sketchbook (cat. 56)

fig. 60 *Greenwich Hospital* c.1831–2 (cat. 52)

'at night under the blaze of the great comet'.[4] This was otherwise known as the Caledonian Comet, which passed across the sky from August to October 1811. That Trimmer, then aged five, remembered it so clearly suggests that it was a significant part of the conversation, and was pointed out to him by his father and Turner. These two recorded observations of astronomical events reveal something of Turner's interest in the sky as being more than the domain of clouds and the weather, and is an attitude that became of growing importance to him in later life.

There is a note in the *Lowther* sketchbook in use c.1810 which reflects the high level of his curiosity about astronomical phenomena. He writes:

Anaxagoras Agarthacus
The Sun 95 millions of miles distant
The Earth
equal to 8 seconds of a degree
The 1200 seconds
The Moon nearly the apparent size of E.[5]

Anaxagoras was the fifth-century-BC Greek philosopher, author of *On Nature*, who discovered the true cause of eclipses. He was the first to show how the cone of the Earth's shadow could eclipse the Moon and vice versa. There are no figures in Anaxagoras; those quoted by Turner are contemporary calculations, and confirm that he knew where to look for astronomical information, and was interested enough to write it down.[6] A later skirmish with a solar eclipse comes in the study made on the back of an envelope (franked 5 June 1824) of a nude figure, perhaps surrounded by dozens of others, gesticu-

fig. 61 *Skyscape over Hawksworth Moor, with Colour Notes* 1816–18, from *Scarborough II* sketchbook (cat. 57)

fig. 62 *Study of the Sky* 1818, from *Skies* sketchbook (cat. 58)

fig. 63 *Sunset – Nancy. Study for 'The Rivers of Europe'* early 1830s (cat. 49)

lating at the eclipsed sun (fig. 58). This is painted on the patch left by the wax seal, and as a whole the work compares with the apocalyptic pictures of John Martin.

Among the possessions listed in the inventory of 47 Queen Anne Street when Turner died were '3 small Telescopes' and 'Pair of Table Globes'. These may have been a terrestrial and a celestial globe.[7] He may have used the telescopes for exploring landscape and seascape as much as the skies, but the presence of these objects is a clue to the nature of his curiosity. On one of his visits to John Fuller in the 1810s Turner climbed Brightling Hill at least to visit the Observatory, if not to use it (fig. 59). Perhaps Fuller accompanied him, for in the drawing, by a gate, is a gentleman clearly waiting for the

artist. Twenty years later, Turner shows an old sailor standing by a telescope looking out over the Thames towards Greenwich (fig. 60). This is in illustration to Rogers's lines, with its purposeful double meaning in reflection of light on water, and mental reflection:

Go, view the splendid domes of Greenwich – Go,
And own what raptures from Reflection flow.
Hail, noblest structures imaged in the wave!
A nation's grateful tribute to the brave.[8]

Turner's primary interest in the sky was to observe and record its partnership with, and effect on, the landscape beneath. There are many instances of his watching the moment-by-moment changes in colour and light in the sky as he sat fishing and at other times. The *Scarborough II* sketchbook has a pencilled skyscape over Hawksworth Moor, with twenty-five or thirty different colour notes written onto it (fig. 61). The most spectacular and extended example is the *Skies* sketchbook (fig. 62)

fig. 64 *Small High Cloud Catching the Sunset (Sunset at Sea)* 1830s (cat. 50)

which appears to be the result of looking at the same patch of sky across three consecutive days. There are in all sixty-five views, the first forty-six pages being the progression from day- or afternoon-light to sunset. What appears to be a stand of deodar trees on p. 42 reappears on p. 52, the apparent second day of the sequence, while on the horizon on p. 10 is a shape that may be the silhouette of Windsor Castle. The castle appears again in pencil studies amongst the group of skyscapes in the third day of the sequence. Other drawings on latter pages in the book are views towards the Castle from Salt Hill, Slough, and yet others, Finberg suggests, are sketches of the fourth of June celebrations at Eton.[9] It seems to me, from research on the ground, that the only

possible place for Turner to have made these sky studies is at the highest point of the slowly rising Salt Hill, outside the Six Bells Inn above St Giles's Church, Stoke Poges, 4½ miles north of Windsor.

There is no evidence of Turner having any serious engagement with the science of meteorology, to the extent, for example, that Constable showed in his cloud studies of the 1820s. Turner's early observations were his own and grew out of his visual research, rather than from talk with scientists. Corroborating Turner's own recollection of his childhood sky-watching, the engraver John Pye recalled that Turner 'never tired of going to Hampstead and would spend hours lying on the Heath studying the effects of atmosphere, and the changes of light and shade, and the gradations required to express them.'[10] There is thus a freshness about the enthusiasm with which he observes, for example, the cold front passing over Coventry in his watercolour of the town for the *England and Wales* series.[11] The suddenness of the change in the weather is vividly described, the town and the left-hand hillside being bathed once more in brilliant light as the storm clouds move away to the right. Turner's instinctive understanding of meteorology in *Coventry* is precise – he has shown how the cold front is embedded with cumulonimbus clouds, which are crowned by a high layer of cirrus.

After the rain, *A Rainbow* (fig. 65). This atmospheric watercolour, with its evidence of a secondary bow to the top right, seems stylistically to belong to the mid-1810s, and may have been made at Farnley, *c*.1816, after a day out on the hills. The subdued colour in the rainbow, practically washed out, indicates that the paper has been soaked and rubbed experimentally, to heighten the effect of rain and sudden light.

In *Arundel Castle with Rainbow* (*c*.1824; fig. 66) the rainbow's evanescence is depicted by Turner's characteristic intense stippling. The refraction of light and the nature

of rainbows was being extensively studied in the 1820s, principally by David Brewster whom Turner came to meet in Edinburgh in 1834.[12] Thornbury writes of the intense arguments between Turner and his Scottish friends on the subject of light and the colours within it, and of Turner's desire to know more about the nature of the spectrum.[13] Within the Edinburgh social circle that Turner entered in 1818, and on visits in 1822 and the 1830s, was Revd James Skene, whose article on Painting in Brewster's *Edinburgh Encyclopedia* includes an observation which gives direct evidence of Turner keeping up with contemporary scientific research: 'aided by the discoveries daily making in the mysteries of light, [Turner's] scrutinising genius seems to tremble on the verge of some new discovery in colour, which may prove of the first importance to art.'[14]

During his 1819 visit to Rome Turner joined the social circle of fellow artists Lawrence, Chantrey and Jackson, the poet Thomas Moore, and Sir Humphry Davy and his wife. The Davys were a popular, well-connected

fig. 65 *A Rainbow with Cattle* ?1816 (cat. 46)

fig. 66 *Arundel Castle with Rainbow* c.1824 (cat. 47)

and well-travelled couple, and although there are no recorded meetings in London between Turner and Davy, and Moore's diary record of Turner in Rome is vague, a meeting in Rome may have been one of many.[15] Davy and Turner had friends and interests in common which might have directed their conversation. They were both passionate about fishing, and well informed about it, and both knew their fish and how and where particular kinds may be caught. Davy's notebooks and correspondence record fishing exploits and experiments on fish, and during the winter of 1827–8, when he was very ill and no longer able to carry out scientific research, he wrote his long treatise, posthumously published, *Salmonia, or Days of Fly-Fishing.*

Like Turner, Davy was also an amateur poet, moved to verse by the beauty of the landscape, animals, birds and art. He wrote encomia on both Byron and his good friend Antonio Canova, praising the latter in the words:

Blameless thy life. Thy manners playful, mild,
Master in Art yet Nature's simplest child.'[16]

Davy's prose descriptions of landscape reveal an eye that could be moved to passion both by geological form, aesthetic quality and colour:

There are quarries of red and yellow marble in various parts of the Apennines between Spoleto and Col. Fiorito, & the walls of Foligno contain pieces approaching to Rosso Antico: but a still finer kind is found in the Marca Bresciana on the borders of the Lago di Garda between Desensano & Pesciara, the stones on the side of the great road many of them rival the Rosso Antico.'[17]

Both Turner and Davy were visionaries, driven to explain and interpret the observable world, and to extrapolate beyond its boundaries. Among Davy's late writings are an extraordinary series of hallucinatory accounts of the rise and fall of civilisations, of invention and change and of interplanetary travel. He expresses his visions in full, exuberant colour, highly tuned and boldly orchestrated across a vast canvas:

I saw below me a surface infinitely diversified, something like that of an immense glacier covered with large columnar masses, which appeared as if formed of glass, and from which were suspended rounded forms of various sizes, which, if they had not been transparent, I might have supposed to be fruit. From what appeared to me to be analogous to masses of bright blue ice, streams of the richest tints of rose-colour or purple burst forth and flowed into basins, forming lakes or seas of the same colour. Looking through the atmosphere towards the heavens I saw brilliant opaque clouds of an azure colour that reflected the light of the sun, which appeared to my eyes an entirely new aspect, and appeared smaller, as if seen through a dense blue mist ... I saw moving round me globes which appeared composed of different kinds of flame but of various colours. In some of these globes I recognised figures which put me in mind of the human countenance, but the resemblance was so awful & unnatural that I endeavoured to withdraw my view from them. 'You are now,' said the Genius, 'in a cometary system; those globes of light surrounding you, are material forms, such as in one of your systems of religious faith have been attributed to seraphs; they live in that element which to you would be destructive; they communicate by powers which would convert your organised frame into ashes; they are now in the height of their enjoyment being about to enter into the blaze of the solar atmosphere. These beings so grand, so glorious, with functions to you incomprehensible once belonged to the earth; their spiritual natures have risen through different stages of planetary life, leaving their dust behind them, carrying with them only their intellectual power.'[18]

These visions were first written down in 1819 (cat. 6), and were prompted, as Davy reveals, by an overwhelming experience that he had had (or had imagined) while sitting alone in the Colosseum in Rome.[19] These were the months preceding Turner's own visit to Rome, and, if nothing else, it reveals a correspondence between the inspiration of the scientist and the palette of the artist.

Davy had a particular purpose to be in Rome and subsequently Naples in the autumn of 1819, precisely the months in which Turner, too, was in both cities. Coincidentally, Davy made the journey across the Apennines from Fano on the Adriatic coast to Rome at almost exactly the same time as Turner. Davy was *en route* between Fano and Rome on 6 and 7 October 1819, and recorded the temperatures by the river Clitumno

and at Nocera in a notebook.[20] Turner must have made the same crossing during the first or second week of October.[21] Davy, as Turner, had specific purposes for this trip, which included: 'to ascertain if anything can be done to preserve the frescoes of Raphael in the Vatican; to determine if the spots in Carrara marble are *iron oxide* & if they can be *obliterated*.'[22]

Patently, Davy's objects were as closely linked to art as they were to science, and we may infer that his undertaking the task of investigating means to preserve Raphael's frescoes, and to get rid of spots in Carrara marble, was a direct result of his being asked to do so by concerned members of the Royal Academy. The atrocious condition of the Raphael frescoes was well known among Royal Academicians. Lawrence wrote to Farington: 'In both the Sistine Chapel and the Rooms of the Raffaele all in too many parts is Ruin & Decay, at least it appears so to me who is not sufficiently prepar'd for the ravages of Neglect & Time.'[23] Davy had a deep-running interest in the arts, which he expressed in 'Parallels between Art and Science', a short essay first published in the *Director* in 1807.[24] Comparing poetry and painting, he wrote: 'Painting ... appeals to the eye by immediate character; it possesses a stronger chain of association with passion; it is a more distinct and energetic language, and acts first by awakening sensation and then ideas.' In the 'truths of the natural sciences', however, he found 'a nearer analogy to the productions of the fine arts' than to the wonders of mechanical invention; and in an illuminating reference to his own attitude to his experimental work he wrote: 'Discrimination and delicacy of sensation, so important in physical research, are other words for taste; and the love of nature is the same passion as the love of the magnificent, the sublime and the beautiful.'

Such sensibilities mark Humphry Davy out as one of the nineteenth century's great connectors. Unwilling – indeed, unable – to see a chasm between the arts and the sciences, he perceived them as integrated parts of a continuum, with nature and the divine as its driving force:

The pleasure derived from great philosophical [i.e. scientific] discoveries is less popular and more limited in its immediate effect, than that derived from the refined arts; but it is more durable and less connected with fashion or caprice. Canvas and wood, and even stone, will decay. The work of a great artist loses all its spirit in a copy ... Nature cannot decay: the language of her interpreters will be the same in all times. It will be an universal tongue, speaking to all countries, and all ages, the excellence of the work, and the wisdom of the Creator.'[25]

Davy combined his love of the ancient world and of painting with his research work on colour chemistry in a pioneering lecture to the Royal Society given in 1815, 'Some Experiments and Observations on the Colours used in Painting by the Ancients.'[26] This grew directly out of studies made in Rome in the autumn of 1814, where he was given access to ancient paintings with the help of Antonio Canova, who, quite apart from being a great sculptor, was the Pope's surveyor of works of art.[27] Not content with investigating the nature of pigments used in antiquity, Davy extended his brief to demonstrate to living artists why some pigments deteriorated over time and others did not. Thomas Phillips read Davy's paper in advance, and exclaimed:

We artists feel a hey-day kind of exhilaration in consequence … It appears that we have better colours than the ancients ever possessed & I believe are better painters.'[28]

When they were in Rome together in 1819, Davy and Turner's interests converged perfectly. Studies in his *Tivoli and Rome* sketchbook[29] reveal Turner to have gone directly to see Raphael's fresco cycle in the Vatican, and to have developed a series of compositional ideas which led ultimately to his Academy exhibit of 1820 *Rome, from the Vatican*. The year 1820 was the three hundredth anniversary of Raphael's death, and then, as now, centenaries had particular resonance. It was thus no coincidence that both Davy and Turner had Raphael on their minds, the former to preserve his work, the latter to rise to the challenge of painting just as brilliantly as he, and to assert Raphael's relevance in the post-Napoleonic world.

Davy's greatest contribution to science was as a chemist and a geologist, two disciplines which in the 1820s closely overlapped. It was his visionary writings, however, his sense of colour and, one might hazard, his conversations that created the common ground between him and Turner. Reaching the end of his active life as a research chemist, Davy's living thoughts transformed themselves into the visions whose poetic intensity predates the similarly impassioned prose of John Ruskin when writing about Turner.

One of Davy's contemporaries and close friends was the mathematician Mary Somerville (fig. 7). She, like Davy, had interests that took her beyond the sciences. Mary Somerville had taken painting lessons in her youth in Edinburgh from Alexander Nasmyth, and she maintained her very real skill as a painter throughout her life (figs. 11, 12). From 1816 until the late 1830s she and her husband, the military doctor William Somerville, lived in London, where they remained at the centre of intellectual and scientific life. There is a reference in Mary Somerville's *Recollections* which touches on the close and enduring friendship that she and her husband had with Turner: 'I frequently went to Turner's studio, and was always welcomed. No one could imagine that so much poetic feeling existed in so rough an exterior.'[30] Turner returned her visits, as William Somerville implied in a remark in which he gives Turner's opinion of a Somerville family portrait.[31] Although no correspondence between Turner and the Somervilles is recorded, there is among Mary Somerville's collection of autographs the last few lines of an undated letter from Turner, snipped off to preserve the signature, which hints at a social arrangement made between them: 'next the 4th of May at ½ past six. Yours most truly, J.M.W. Turner. 47 Queen Ann St, Sunday Eg.'[32] There are no Turners in the only known list of Mary Somerville's collection of pictures,[33] but in 1997 the author discovered two pen and ink drawings by Turner among the folio of Mary Somerville's watercolour and gouache paintings in a family collection. These are both preliminary studies for *Liber Studiorum* subjects, *Basle and Bridge* and *Goats*,[34] and can only reasonably have descended to their present owner if they had been given by Turner to Mary Somerville, or if (less likely) she had bought them from him. Turner gave his pictures away only rarely, and if he gave these to Mary Somerville it will have been in recognition of their evident mutual esteem and respect.

Somerville's first major published work was *Mechanism of the Heavens* (1831). Turner owned few books on scientific subjects,[35] but he did have a copy of the first edition of this book, suggesting that either Mary Somerville gave him one, or that (less likely) he was interested enough in the subject to buy his own. The bulk of the book is densely written mathematics, but in the chapter 'Progress of Astronomy' the author discusses Galileo's life and achievement, and as John

fig. 67 *Study for 'Galileo's Villa' c.*1826–7 (cat. 51)

Gage has pointed out this may have been enough to prompt Turner to incorporate something of Somerville's ideas into his work.[36] An early study for *Galileo's Villa* (fig. 67), has within it a lightly sketched diagram of the solar system. The text which this study illustrates comes from *Italy* (1830), the collection of poems by Samuel Rogers who was one of the catalysts of the social, intellectual and conversational circle that included Turner and the Somervilles:

   Sacred be
His villa (justly was it called The Gem!)
Sacred the lawn, where many a cypress threw
Its length of shadow, while he watched the stars!
Sacred the vineyard, where, while yet his sight

Glimmered, at blush of morn he dressed his vines
Chanting aloud in gaiety of heart
Some verse of Ariosto![37]

The planetary diagram in *Galileo's Villa* returns again in a more considered manner in a subsequent vignette, *Mustering of the Warrior Angels*, made for Milton's *Poetical Works* published in 1835 (fig. 68).

In *Mechanism of the Heavens* Mary Somerville conjures up imagery that would have been deeply impressive to an enquiring and flexible mind like Turner's:

The heavens afford the most sublime subject of study which can be derived from science: the magnitude and splendour of the objects, the inconceivable rapidity with which they move, and the enormous distances between them, impress the mind with some notion of the energy that maintains them in their motions with a durability to which we can see no limits.'[38]

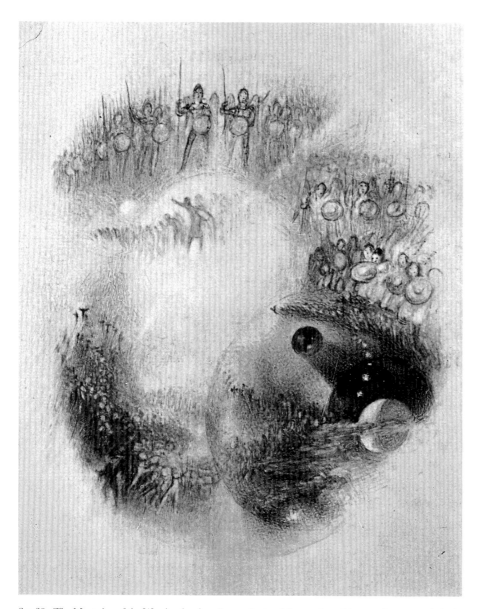

fig. 68 *The Mustering of the Warrior Angels* c.1833–4 (cat. 53) (reproduced larger than actual size)

fig. 69 *Imaginative Sunset* mid 1830s (cat. 54)

Abstract thoughts of this kind, published and no doubt discussed also in conversation, are of a kind that propel us directly from such tight little figurative vignettes as those for Rogers and Milton to some of Turner's late paintings, particularly *Light and Colour (Goethe's Theory)* (exh. 1843; Tate Gallery) and *The Angel Standing in the Sun* (exh. 1846; Tate Gallery).

One extraordinary painting that Turner first exhibited in 1834, and then tampered with and, unusually, exhibited again in 1839, is *The Fountain of Indolence*. On its second appearance, Turner renamed it *The Fountain of Fallacy* (fig. 70). In its first incarnation it is likely that the painting's inspiration came from the poetry of James Thomson, a life-long favourite of Turner's.

Thomson's *The Castle of Indolence* describes 'a fountain of Nepenthe rare'

Whence, as Dan Homer sings, huge pleasaunce grew,
And sweet oblivion of vile earthly care,
Fair gladsome waking thoughts, and joyous dreams more fair.[39]

'Nepenthe' is a species of pitcher plant, having sedative properties; so there is little wonder that the naked, happy figures are crowding to drink from its fountain.

Gage suggests that the change in the painting's title reflects Mary Somerville's influence on Turner. This is plausible, given a passage Gage quotes from Somerville's second book, *On the Connexion of the Physical Sciences* (1834):

A consciousness of the fallacy of our judgement is one of the most important consequences of the study of nature.

This study teaches us that no object is seen by us in its true place, owing to abberation; that the colours of substances are solely the effects of the action of matter on light, and that light itself, as well as heat or sound, are not real beings, but mere modes of action communicated to our perceptions by the nerves.'[40]

The original existence of the painting as *The Fountain of Indolence*, however, owes as much to the heady visions of Humphry Davy, who had seen inhabited bubbles rushing through space, as to James Thomson's poem. In addition to the passage quoted above, Davy described in *Consolations in Travel* abundant vistas, teeming cities, and the growth, maturity and death of civilisations:

I saw a great extent of cultivated plains, large cities on the sea shore, palaces, forums and temples ornamenting them; men associated in groups, mounted on horses, and performing military exercises; galleys moved by oars on the ocean; roads intersecting the country covered with travellers and containing carriages moved by men or horses.[41]

This is vision of a Turnerian kind: the imagery of *Dido and Aeneas* rolled into the endless landscape of *Raby Castle*. We do not know how attentive Davy had been to Royal Academy exhibitions, but as a friend of the President Sir Thomas Lawrence,[42] Chantrey and Wilkie, he cannot have been blind to them, nor untouched by the ubiquity of Turner's production.

Whatever it was that prompted Turner to amend his *Fountain* painting, and rename it *The Fountain of Fallacy* prompted him also to add four lines from his extended poem *Fallacies of Hope* to the painting's entry in the 1839 Academy catalogue:

Its Rainbow-dew diffused fell on each anxious lip,
Working wild fantasy, imagining;
First, Science in the immeasurable Abyss of thought
Measured her orbit slumbering.

These lines read as if they are an extract from a larger poem; the fact is that unless there was once some extended draft that is lost, each snatch of verse in *Fallacies of Hope* stands alone. Thus the implication of preceding text in the first line is entirely misleading, so all we can squeeze from these opaque lines is, as John Gage put it, that 'science, or thought, is capable of measuring, although it is itself immeasurable'.[43]

Thoughts of the sky and the planets gave Turner a key to the infinity that he hinted at in his lines from

Fallacies of Hope. He was moved even in his dreams by the pioneering balloon flight made in 1836 by Monck Mason, Charles Green and Robert Hollond MP. He must have met Hollond at a gathering soon after the flight, for Turner wrote to him urging him to make a painting of what he had seen: 'Your Excursion so occupied my mind that I dreamt of it, and I do hope you will hold to your intention of making the drawing, with all the forms and colours of your recollection.'[44] Turner's eagerness to talk about and experience the excitement of flight, and the new perspectives it offered, extended even as far as his encouraging an indifferent amateur artist to show him his drawings. There is a developing sense of elevation in Turner's paintings of the late 1830s, which may have some relation to his chance remark about flight. The viewer's feet are high off the ground in *Juliet and her Nurse* (exh. 1836), and in *Rome, from Mount Aventine* (exh. 1836) and *Cicero at his Villa* (exh. 1839) the viewpoint is high above the foreground detail. This is in marked contrast to paintings of the first quarter century of Turner's career, where the viewer stands solidly on *terra firma* addressing the scene and being part of it. The aerial perspective in the later paintings has the effect of tipping the horizon down, extending the vista and lowering distant features. In these paintings of the 1830s the viewer is part of, and addresses, the sky.

Turner returned, like a voyaging comet, to the imagery of the planetary system in his final years. His last known interpretation was recalled by the history painter Solomon Hart (1806–81) whose painting *Galileo, when Imprisoned by the Inquisition at Florence, Visited by Milton* was hanging in the Academy on Varnishing Day 1847. Turner, without the artist's knowledge, added in chalk a diagram of the solar system as if it were a halo hanging above Galileo's head.

Turner was upon the point of effacing his addition, but [Clarkson] Stanfield, who was much interested, hastened to me, to persuade me to preserve the lines. He mixed up some paint and stood over me whilst I secured them with colour. All thought that Turner's suggestion had much improved my picture.[45]

This says less about Turner's interest in the solar system than about his concern that a fellow painter's work should be seen at its best.

fig. 70 *The Fountain of Fallacy ('The Fountain of Indolence')* 1834 oil on canvas *The Beaverbrook Foundation. The Beaverbrook Art Gallery, Fredericton, New Brunswick*

5

FROM SAIL TO STEAM: THE ABSENCE OF TROUBLE

Turner provided the classic image of the painful but inevitable change in early-nineteenth century shipping from sail to steam. *The Fighting Temeraire Tugged to her Last Berth to Be Broken Up, 1838* (exh. 1839; National Gallery) gives equal prominence both to sail and to steam, and casts both ships in heroic roles – the large, slow, quiet, pale wooden ghost is tugged to the breakers by a noisy little nippy iron workhorse with another job to do tomorrow.[1] Times change; life moves on.

As an indicator of his general attitude to the technological progress that he saw all around him, we should note at the outset that although there are many shipwrecks in Turner's paintings they are all of sailing ships

or rowing boats. There are no wrecked steamships in Turner. This runs against the pattern of events – about 1,400 steamships have been traced as active in British waters up to 1840,[2] and among these boilers blew up, ships caught fire, paddle-wheels shattered. Turner will have known about all that, and perhaps experienced accidents, but the absence of trouble in his depiction of the steamship directs us to the view that Turner regarded steam as a manifestation of security, human endeavour and hope. Ruskin, however, reacted pessimistically to the silent message of Turner's steamships, and most later critics have followed him. Like thousands of his contemporaries Turner travelled by steam often

fig. 71 *Falmouth Bay* 1811, from *Ivybridge to Penzance* sketchbook (cat. 75)

(*right, top*) fig. 72 *River Scene with a Smoky Steamboat* c.1815–17, from *Walmer Ferry* sketchbook (cat. 76)

(*right, bottom*) fig. 73 *Steamer Leaving Harbour* mid 1840s, from dismembered *Whalers* sketchbook (cat. 72)

fig. 74 *Dartmouth on the River Dart* 1822 (cat. 62)

and happily enough, both over the water and on the rails, backwards and forwards from Blackfriars to Margate, around the Kent coast and across the Channel. 'I must go by steam,' he wrote, probably in the 1840s, to David Roberts.[3]

As early as the 1820s steam travel by sea had come to stay, a certainty that Humphry Davy had already expressed: 'The dominion of Britons in Asia may share the fate of that of Tamerlane or Zengiskhan; but the steam-boat which ascends the Delaware or the St. Laurence will be continued to be used and will carry the civilization of an improved people into the deserts of North America and into the wilds of Canada.'[4] Exchanges in letters among Turner's friends show how the wonders and excitement of steam impressed even the most sophisticated of artistic people: Thomas Donaldson told that he had been on the quickest passage

ever made by steamship from London to Edinburgh in 1824 – forty-three hours – and Thomas Phillips RA, writing to Dawson Turner, said that the steamboat had shortened the journey time between Liverpool and Dublin by three hours.[5]

Turner's experience of steamship construction, however, was not as modern as his consideration of the *idea* of the steamship. Almost every little harbour around the British coast had its boatyard, and on his travels Turner had seen dozens with sailing boats on the stocks (fig. 74). But although he could have sought them out, there seems to be no trace in the sketchbooks of steamship yards, with their heavy cranes and engine-works, and there are none in the oils or watercolours. This suggests that by the time steamships came along, Turner's self-imposed mission to define an identity for Britain in

(*right, top*) fig. 75 *Dover* c.1825 (cat. 63)

(*right, bottom*) fig. 76 J. T. Willmore, after J. M. W. Turner, *Dover* 1851 (cat. 64)

Picturesque Views of England and Wales and his other series of watercolours for engraving had matured into a wider view. He showed no interest in how steamships were made, but instead used them as the transport for a moral, symbolic and cultural burden.

There are four finished watercolours of Dover harbour in the *England and Wales* cycle, and two more of the coast nearby.[6] Dover, the nearest British port to the Continent, had always had strong symbolic associations as Britain's premier gateway and bastion, and it is probably no coincidence that of all the coastal ports in the *England and Wales* series it is only in his Dover pictures that Turner incorporates a steamship. The modernism of the steamships underscores Turner's view of the patriotic importance of Dover in Regency Britain. The steamer in *Dover* (fig. 75) steams cheerily out to sea, while all around it sailing ships do the wind's will, and oars-

men puff and pant. Another work, *Dover Castle* (1822; Museum of Fine Arts, Boston – engraving, called *Dover*, fig. 76), has a more complex imagery: here the steamer, converted from a sailing boat and loaded to the gunwales with passengers, bustles past a stranded brig, while two small sailing boats find themselves in some kind of difficulty in the choppy sea. The meaning is straight forward enough – the steamer is as well able to withstand angry seas as Dover Castle, high above it, has stood guard for Britain for a thousand years. Turner expected his audience to understand the language of these pictures with ease, and indeed most men and women with experience of the sea could read them then as well as we might now read a picture of a dangerous traffic situation on a motorway.

Steamships gradually made marine travel safer, quicker and possible to timetable reliably. They also

fig. 77 *Between Quilleboeuf and Villequier* c.1832 watercolour (TB CCLIX 104)

fig. 78 *? Ehrenbreitstein from Coblenz c.*1841 (cat. 69)

made it dirtier, and might have been more accurately named 'smokeships'. Coastal journeys began from countless piers in London, and by the 1830s steamships had become a constant presence on the Thames and around the coasts. They also revolutionised travel on German and French rivers, particularly the Seine which meanders like a mountain road, doubling back on itself again and again between Le Havre and Paris, and running between hilly banks, over shoals, sand bars and weirs. These conditions made sailing ships a slow and expensive mode of transport, and hampered trade. Steamships, on the other hand, could negotiate these difficulties with relative ease and, as Turner demonstrated in his watercolour *Between Quilleboeuf and Villequier* (*c.*1832; fig. 77), were used to tug sailing ships out of difficulty.[7] Steamboats were also a common sight on the Rhine from the 1820s, and for this reason they became

a natural part of the iconography in Turner's views on the Seine and the Rhine (figs. 78–80). Because of the length of French and German rivers, and the distance inland of many of the great cities, the steamboat made an impact on the continental riverscape to an extent that never occurred in England. Thus, in Turner's British subjects, the steamboat is a machine for the high seas and the estuaries, while in his continental subjects it lives the more tranquil life of the inland river. That is the nature of the divide between the two characters of the steamboat in Turner, and it allows him ample scope to develop its metaphorical richness.

Despite its title, the prominent feature in *Staffa, Fingal's Cave* (exh. 1832; fig. 81) is the steamship which rounds the island. The cave itself is indistinct. The weather and sea conditions apparent in the painting do not follow Turner's recollection of them, as written to James Lenox in New York, fourteen years after the event: 'a strong wind and head sea … rainy and bad-

79

(*above*) fig. 79
Honfleur 1830s
(cat. 70)

(*left*) fig. 80
James B. Allen,
after
J.M.W. Turner,
Caudebec 1834
(cat. 71)

(*right*) fig. 81
Staffa, Fingal's Cave
exh. 1832
oil on canvas
*Yale Center for
British Art, Paul
Mellon Collection*

looking night coming on … The sun getting towards the horizon, burst through the rain-cloud, angry, and for wind; and so it proved, for we were driven for shelter into Loch Ulver.'[8] The sea Turner depicts in the painting is practically a flat calm, with seabirds swooping about and low waves transmogrifying themselves at the left into basaltic rock. The island is being revealed in an opalescent light as the clouds pull back and the sun sets within a halo, a phenomenon that forecasts rain. If Turner remembered the sea conditions aright in his letter, the painting is of a different kind of day altogether, an amalgam of experiences brought together to create a purposeful allegory. It was exhibited with lines from Walter Scott which introduce the powers of nature:

Nor of a theme less solemn tells
That mighty surge that ebbs and swells,
And still, between each awful pause,
From the high vault an answer draws.'[9]

Scott's lines, however, have a context in which the poet compares the natural forms of Staffa with human architecture:

Nature herself, it seem'd would raise
A Minster to her Maker's praise!'

The verse concludes:

Nor doth its entrance front in vain
To old Iona's holy fane,
That Nature's voice might seem to say,
'Well hast thou done, frail Child of clay!
Thy humble powers that stately shrine
Task'd high and hard – but witness mine!'

fig. 82 Robert Carrick, after J.M.W. Turner, *Rockets and Blue Lights* 1852 (cat. 74)

'Fane' here means 'temple'. Turner's introduction of the character of the stout-hearted man-made steamship in contrast to the grandeur of nature is a direct parallel to Scott's passage taken as a whole. Where the clouds move apart, the smoke from the steamship rises and falls, then arcs upwards turning from dusky black to pure white as it reaches the higher air. By quoting part of the text, but not the manifestly relevant whole text, in his catalogue entry, Turner's main point remains in the image, to be underscored by a fuller reading of Scott. We do not know how many lines of poetry Turner was 'allowed' by the Academy catalogue editor. Not so many, perhaps; but before accepting, as many critics have, that the steamship is in imminent trouble, we should reflect on

the approving tone of Nature's voice, as relayed by Scott: 'Well hast thou done, frail Child of Clay!'

Turner's other great steamship oil subject is *Snow Storm – Steam-Boat off a Harbour's Mouth …* (fig. 129). This will be looked at in fuller detail in Chapter 7, but in the meantime there are details in it to suggest that, although in a very tricky situation at sea, it is not alone with the elements, nor far from its haven. There is a suggestion of the presence of another steamship beyond and perhaps of groynes or some kind of sea wall on the right. Though almost swamped, the steamer is paddling away bravely, and, as the full title indicates, is methodically '*Making Signals in Shallow Water, and Going by the Lead*'. 'Going by the lead' means taking soundings in shallow water using a lump of lead on the end of the rope. In closely following the rules of the sea in extreme weather, the crew is navigating the ship in a seamanlike manner with every likelihood of a safe landfall. This is the attitude that Turner's painting reflects, and that his title for it tells us. He might have said 'going by the *rules*.'

A writer in the *Quarterly Review*, discussing the new edition of Thomas Campbell's *Poetical Works* (1836), identified Turner and Campbell as together having pioneered the expression of 'a new object of admiration, – a new instance of the beautiful, – the upright and indomitable march of the self-impelling steam-boat,'[10] and quoted lines from Campbell's *On the View from St Leonard's.*[11] In making this judgement, the reviewer was thinking specifically of Turner's *Annual Tours – Wanderings by the Seine*, which was published as text and engravings in two parts in 1834 and 1835. Evoking *Quilleboeuf to Villequier*, the reviewer firmly associated Turner with his subjects, and observed: 'The tall black chimney, the black hull, and the long wreath of

smoke left lying on the air, present, on *his* river, an image of life, and of majestic life, which appears only to have assumed its rightful position when seen amongst the simple and grand productions of nature.'

Though there are many steamships in Turner's works on paper, they make relatively rare appearances in his oil paintings by comparison with the extent with which he painted sailing vessels. Apart from the three exhibited oils mentioned above, in only two others – *Peace, Burial at Sea* (exh. 1842; Tate Gallery) and *Rockets and Blue Lights* (exh. 1840; Sterling and Francine Clark Art Institute, Williamstown; chromolithograph, fig. 82) – can steamships be said to be prominent, so we must keep a sense of proportion about their place in his *oeuvre*. In the other completed oils, *The Chain Pier, Brighton* (1828–9; fig. 100), *Wreckers – Coast of Northumberland, with a Steam-*

fig. 83 *Wreckers – Coast of Northumberland* exh. 1834 oil on canvas *Yale Center for British Art, Paul Mellon Collection*

Boat Assisting a Ship off Shore (exh. 1834; fig. 83), *Fishing Boats with Hucksters Bargaining for Fish* (exh. 1838; Art Institute of Chicago) and *The New Moon* (exh. 1840; Tate Gallery), the steamer is a background presence, a silent reminder of encroaching and, in the case of *Wreckers*, beneficial modernism. In *Wreckers* the steamer lies offshore waiting to give aid, as the title makes clear, to a sailing ship in trouble before it falls into the hands of human vultures.[12] Thus, the new technology is quite literally a ghostly *deus ex machina*, frustrating for once the power of the people luring a sailing ship aground.

Margate developed rapidly as a popular seaside destination for Londoners during the 1820s and 1830s. Visitors came in their greatest numbers by steamship, landing both at the Pier, a stone harbour wall built by John Rennie (opened 1815), and Jarvis's Landing Place, the nearby jetty which came into service nine years later.

Mrs Booth's house, where Turner stayed regularly in the 1830s and 1840s, was on high ground in sight of both the Pier and jetty, and it was from here, and the foreshore immediately below the house, that Turner made many of his sketches of steamship subjects. Figs. 84 and 85 and many others are taken directly from this point, and reflect Turner's fascination with the marine and human activity there. These Margate studies come closest in their general character to studies of the French and German rivers (fig. 78). There is the same bustling, peaceable activity, made the more appealing to him through familiarity and enjoyment.

Turner's engagement with steamships, and his depiction of lighthouses and modern maritime inventions can all be seen as manifestations of his excitement about and support for devices that would make seafaring safer. Four new devices, all invented by the same man, were

fig. 85 *Steamer with Yellow Smoke, Margate* 1840s (cat. 73)

(*left*) fig. 84 *Harbour with Figures and Shipping, Margate c.*1826 (cat. 65)

fig. 86 *Life-Boat and Manby Apparatus* 1831 (cat. 68)

fig. 87 Manby
Apparatus Mortar
(cat. 77)

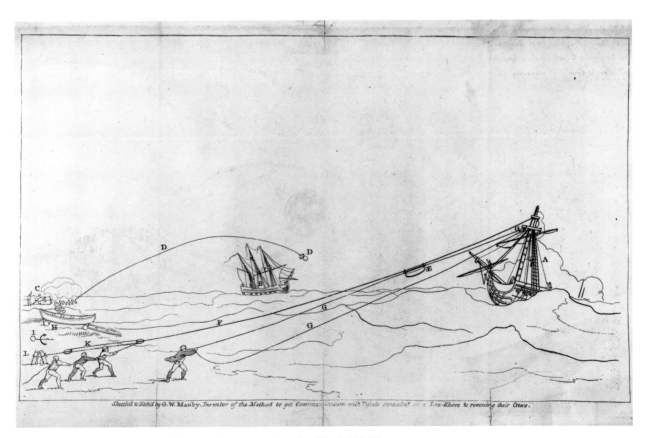

fig. 88 G.W. Manby, Diagram of Manby Apparatus in action *The British Library*

celebrated together in *Life-Boat and Manby Apparatus Going Off to a Stranded Vessel Making Signal (Blue Lights) of Distress* (exh. 1831; fig. 86). Captain George William Manby (1765–1854) was a Great Yarmouth barrack-master and inventor who was convinced that many of the lives lost in coastal shipwrecks could be saved. Since the early 1810s he had been perfecting a method whereby a small cannonball, with a rope attached to it, could be fired from the shore into the rigging of a ship in difficulty, thus establishing a line for the sailors to secure, cling to and escape (figs. 87–8). Manby gave an early demonstration of the apparatus to 'a multitude' beside the Serpentine in London in 1812, by firing a line across the lake into a tree. The Duke of York was present, as were 'some of our principal chemists and people of distinction … The neatness and obvious utility of this experiment excited great admiration.' At the same event, Manby

demonstrated his lightball for shipwrecks at night, which, as the *Times* reported, 'was thrown up to a considerable height, burst at its highest elevation, and poured down a shower of stars, which, in the darkness of a tempestuous night, must have thrown powerful illumination upon its subject'.[13] Two years later, Manby had devised a way of making the shotcarrying the rope into a lightball, thus incorporating the two inventions. The shot was drilled with four holes 'to receive a like number of fuses … filled with the fiercest and most glaring composition, which when inflamed at the discharge of the piece, displayed so splendidan illumination of the rope, that its flight could not be mistaken.'[14]

For twenty years Manby tried to interest successive British and French governments and royalty in his ideas and, although constantly frustrated, his persevering nature drove him on. By 1823, when a government

(*above*) fig. 90 *The Bell Rock Lighthouse* 1819 (cat. 61)

(*left*) fig. 89 *Firing Rockets off the Coast* mid 1820s (cat. 67)

select committee report on Manby's Apparatus was published, sets had been sited around the Norfolk and Suffolk coast, where they had contributed to the saving of more than two hundred lives (figs. 87, 88). But despite its achievements, full official acceptance of the Apparatus was constantly subject to bureaucratic obfuscation and delay, prompting the *Edinburgh Review* to declare in 1823 that 'we deem it important to spread as widely as possible the knowledge of the plan and to point out the means of effecting its general adoption.'[15] Manby did not give up. He lectured widely around the country, coming in 1830 to the Royal Institution where Faraday gave him a platform.[16] Among Manby's other inventions was a way to make any ordinary rowing boat unsinkable, by lashing six or seven empty air-tight casks under the gunwales. Inventive thinking made it clear to him that such an adapted boat – a lifeboat – could be used to save lives at sea. In a further embellishment he devised

an anchor and rope arrangement to haul lifeboats back to the shore through the surf.

Turner's painting incorporates all four of these inventions: the lightball carrying the line is throwing out flames directly above the distress flare of the stranded vessel, while a lifeboat, attached by a rope to the anchor on the shore, is rowed out to rescue the seamen. As an indication of the rapidity of the whole operation, Turner has shown the cloud of smoke from the mortar blowing away directly above the anchor on the beach.

Turner and Manby had a mutual friend in the Yarmouth banker, collector and amateur scientist Dawson Turner (1775–1858). Dawson Turner underwrote Manby's experiments for many years, and the pair developed a relationship which is thoroughly documented, in all its heats and colds, in the volumes of the Dawson Turner Correspondence in Trinity College Library, Cambridge.[17] There is, however, no surviving evidence that this mutual friendship prompted J.M.W. Turner to take Manby's inventions as a subject, though it is entirely possible. Turner will have been

fig. 91 *Lighthouse against a Stormy Sky* mid 1820s (cat. 66)

as aware as any thoughtful Englishman of the public campaigns to install Manby's Apparatus, and this would have been impetus enough for him to choose it as his subject. There is, however, one further connection that may not be entirely coincidental. On 12 May 1831, when Turner's *Life-Boat and Manby Apparatus* was hanging in the Royal Academy, Captain George William Manby himself was elected a Fellow of the Royal Society at a meeting in Somerset House, only yards from the room in which Turner's painting hung.[18] If the painting was Turner's deliberate canvass for Manby's election, it was eloquent indeed.

During the years of Turner's maturity the technology and design of lighthouses improved rapidly, until by 1830 there were as many as eighty-one at work around the British and Irish coasts.[19] Their design attracted some of the most brilliant engineers and inventors, and indeed Captain Manby invented a device by which lighthouses could identify themselves to seamen through a signal particular to each. The Bell Rock Lighthouse,

off the east coast of Scotland twelve miles south east of Arbroath, was completed in 1811 to the designs of Robert Stevenson (1772–1850). In the latter years of the 1810s Stevenson wrote his *Account of the Bell Rock Light House* and in 1819 approached Turner to make a drawing of the lighthouse for engraving as an illustration to his *Account* (fig. 90). The commission seems to have taken less than two months from beginning to end – from May until early July 1819[20] – and as there was no question of Turner visiting the lighthouse to make the drawing, he carried it out from other men's sketches. All Turner needed to make the drawing was a more or less technical view of the structure; the rest came from his imagination and memory of sea conditions. The watercolour and its engraving, made in 1824 by John Horsburgh, is a study of tempestuous weather with thunder and lightning, causing the waves to wrap around the lighthouse like fingers, and spray to reach 100 feet up to the lightroom.

Unlike his earlier lighthouse subject *The Eddystone Light House* (*c*.1817; whereabouts unknown – engraved by T. Lupton 1824), where recent wreckage floats in front of the tower, *The Bell Rock Lighthouse* contains no suggestion that it might fail in its duty, as the presence of two ships keeping well away amply demonstrates. This was Turner's most dramatic and unequivocal rendering of a lighthouse to date, a clear demonstration of his feeling for the stoutness of Stevenson's structure, and his conviction for the boundlessness of man's ingenuity. Although Turner had drawn and painted lighthouses before,[21] he appears to have paid more attention to the subject after the Bell Rock commission. There are lighthouse studies in the *Folkestone* sketchbook of 1821,[22] and principal subjects include *Longships Lighthouse, Land's End* (*c.*1834; J. Paul Getty Museum, Malibu, California) and *Orfordness* (*c.*1827) for which fig. 91 is a study.

The year after he exhibited *The Fighting Temeraire*, Turner showed a pair of paintings that, when seen together, speak just as eloquently as that painting of the passing of sail and the coming of steam. Unlike *The Fighting Temeraire*, however, circumstances have conspired to distort the antiphonal voices of *Slavers* and *Rockets and Blue Lights*. They were exhibited at the 1840 Royal Academy exhibition in 'conspicuous places' according to Revd John Eagles, writing in *Blackwood's Magazine*.[23] This does not in itself suggest that they were hung side by side, but that they were intended to be pendants is indicated by a number of factors. Firstly, they are painted in opposing colour values, the former principally red and yellow, the latter blue and yellow; this was a manner of colour handling that Turner practised in acknowledged pendants during the late 1830s and 1840s, for example, *Modern Italy* and *Ancient Italy* (1838), *Peace* and *War* (both 1842). Secondly, they were painted over the same few months and exhibited together on the same size canvases. Thirdly, they remained together as a pair with Turner's dealer Thomas Griffiths, when Ruskin advised the collector William Wethered to buy both.[24]

In the event Wethered bought neither and the paintings went their separate ways. *Slavers Throwing Overboard the Dead and Dying – Typhon Coming on* (Museum of Fine Arts, Boston) takes for its subject a criminal and inhuman act, the throwing overboard from a sailing ship of slaves during a storm, an outcry against the barbaric practice of slavery, already banned in Britain, and soon to be prohibited elsewhere in Europe. On the other hand, *Rockets and Blue Lights (Close at Hand) to Warn Steam-Boats of Shoal-Water* (Clarke Art Institute, Williamstown; chromolithograph fig. 82) describes an event that was becoming increasingly common around the shores of Britain. Using new rocket technology, the new steamships are warned by concerned and organised onlookers to keep well away from the shallows and thus save themselves and their cargoes. The one, using shock and allegory, shows the final kick of barbarism at sea, whilst the other reflects upon new maritime safety measures successfully deployed. Despite the vicissitudes of the past 120 years, the two paintings refuse to be parted completely. They might have gone anywhere, but have settled only 120 miles apart in public collections in Massachussetts. Read as a pair, their meaning is enhanced: steam, in Turner's peaceable world-view, has finally taken over from sail.

INDUSTRY AND CONSTRUCTION AFTER WATERLOO 1815–51

In the 1815 Royal Academy Turner exhibited *Crossing the Brook* (fig. 92), in which two young girls and their dog make a short journey. Both Jack Lindsay and Eric Shanes have shown this to be Turner's symbolic assertion of girls passing from childhood to adulthood.[1] The panorama in the distance, however, a generalised view of the Devon landscape with Calstock Bridge over the River Tamar in the middle ground and Dartmoor rolling away in the distance, takes the meaning of the painting further, from human universals to national specifics. Just beyond the bridge are the arching lines of a huge water-wheel, with cascades curving from one aqueduct to another. Big wheels had gradually intruded into the Devon and Cornwall landscape as tin- and copper-mining developed,[2] and by showing industrial works within a luxurious Claudean landscape, Turner is reflecting in *Crossing the Brook* on a nation making its own crossing in peacetime to an industrial economy.

The Academy exhibition opened as usual in 1815 in late April. It was Turner's practice to work up his large exhibition pictures during the preceding autumn and winter, but this year, between the painting of the picture and its exhibition, there was an entirely unforeseen but climactic event. Napoleon, safely imprisoned on Elba since April 1814, escaped in March 1815 and made a triumphal return to France. *Crossing the Brook* had, however, been conceived and painted during the first months of peace after twenty-two years of war, providing purpose to its mood of optimism, and its graceful description of parallel change within woman and within the nation. Turner's apparently bizarre timing, which has clouded the full meaning of the picture, is explained by the fact that when it was exhibited the world had changed again, and Britain, her allies and France were squaring up for their final battle.

Change is the leitmotif of many of Turner's industrial subjects in the years around 1815. *The Vale of Ashburnham, The Vale of Heathfield* and *Crowhurst* (Chapter 3; figs. 48–50) reflect on the passing of an out-moded industry; whilst

Leeds (1816; fig. 93; sketchbook study fig. 94) looks at change from the other end. Turner shows Leeds in a wide panorama, its factories hard at work and chimneys pouring out smoke. In the foreground a cloth-worker is carrying a bundle of newly woven cloth to be dried by the tentermen in the sun; labourers build or repair a stone wall; and a pair of milk-carriers trudge up the hill on their donkeys.[3] There is a powerful buzz about this picture that is all too absent in *Crowhurst*. Leeds was enjoying an economic boom around 1815. New mills had been built in the first decade of the century to supply the national market for all kinds of cloths and rope products. Two of these mills, Marshall's (left), and Benyon's (centre) are prominent in the picture, as indeed is the evidence of Leeds encroaching on the countryside as its prosperity and population increases. Turner was a regular visitor to Leeds, passing through the town on his way to Farnley, near Otley, where he spent weeks at a time in the 1810s and early 1820s with the landowner and politician Walter Fawkes and his family.[4]

In August 1822 Turner travelled by sea from London to Edinburgh, to witness the state visit to Scotland of King George IV. As his ship travelled up the east coast, he made dozens of drawings of coastal horizons and details in the 'King's Visit to Scotland' sketchbook.[5] Almost certainly during a halt on this journey Turner made the pencil drawing *The Hurries* (fig. 95).[6] This is a rare example of the under-drawing of a painting surviving, because Turner did not in the event obliterate it with watercolour. It contains the kind of intricate and well-observed detail that Turner delighted in, and in this is comparable to his Fonthill drawings of 1799 (fig. 18).

Shields, on the River Tyne (fig. 96), later engraved for *The Rivers of England* series, is also a recollection of a stop-over on the 1822 voyage. It harks back in its Rembrandtian mood to *Limekiln at Coalbrookdale* of the 1790s (fig. 36) but, unlike the early work, has a sense both of specific place and of real time. Coal was delivered

fig. 92 *Crossing the Brook* exh. 1815 (cat. 78)

fig. 93 *Leeds* 1816 watercolour heightened with bodycolour over pencil *Yale Center for British Art, Paul Mellon Collection*

fig. 94 *Leeds* 1815–16, from *Devonshire Rivers no.3 and Wharfedale* sketchbook (cat. 79)

fig. 95 *'The Hurries' – Coal Boats Loading, North Shields* 1822
(cat. 80)

from the Northumberland and Durham mines by
barges, or 'keels', which for decades had plied up and
down the River Tyne taking the coal to colliers for
onward transport by sea to London. From the early
1820s the keelmen were facing competition from railway
lines built to take coal from the mines, and high above
the loaded keels Turner shows a railway wagon ready to
discharge its coal into a collier below. Twelve years later
Turner came back to the subject and composition in the
oil *Keelmen Heaving in Coals by Moonlight* (exh. 1835; fig. 97).
The central gap of the river in *Shields* is widened in
the later oil painting, and Turner has omitted the
poignant detail of the railway wagons. Instead, the
subject is generalised, as was *Limekiln at Coalbrookdale*.
An inference that may be drawn from this is that the
1823 watercolour is, like *The Hurries*, nearer to reportage,
perhaps even affected in response to casual conversations on the Shields quayside, and from information
picked up on the spot during his 1822 sea voyage.

As we know from the many written and drawn notes
throughout his sketchbooks – for example the details of
sailors in the *Nelson* sketchbook[7] – Turner was avid in
his search for information, and tended to talk to people
as well as to explore with his eyes. Twelve years on,
however, Turner's interest in the subject had shifted
to its value as a source of poetic evocation in oil of
marine industry, and as one of a pair of commissioned
pendants reflecting two maritime nations, one, Venice,
in decline through decadence and the other, Britain, in
the ascendancy.[8]

Turner's subject matter embraced the extraordinary
and the everyday, and through his genius and powers of
selection he made the everyday appear extraordinary.
In *Kirkstall Lock, on the River Aire* (fig. 98), the slow, placid,
but soon-to-be-outdated mode of commercial transport,
the canal barge, is pincered between two instances of
forceful physical activity – the dashing coach from
Bradford on the right, and the determined construction
work in the expanding town of Leeds on the left. The
serpentine line of the Bradford road crosses the canal
bridge in the middle distance and disappears westwards

fig. 96 *Shields, on the River Tyne* 1823 (cat. 81)

into the late afternoon sun. With perfectly ordinary pictorial ingredients, the staple of any topographer, Turner emphasises how times are changing and how, through laying strong foundations, Britons were harnessing their land. It is no coincidence that more works in this exhibition are taken from *The Rivers of England* than from any other of Turner's sets of watercolours for engraving (figs. 66, 74, 96, 98). It indicates that this series reflects particular thinking, that Turner saw England's rivers as work places, the focuses of industrial activity and equipment across the country. Out of seventeen *Rivers* subjects, ten are shipping subjects, two show canal locks and one shows logging.

The 3rd Earl of Egremont (fig. 2) used his huge fortune and philanthropic nature to make innovations in agriculture, construction and technology.[9] Some of this was reflected in two of the subjects Turner painted for

the dining room at Petworth. *Chichester Canal* and *The Chain Pier, Brighton* (the Petworth version is known as *Brighton from the Sea*; in place by 1829; figs. 99, 100) recall construction projects which Lord Egremont was closely involved with in the 1820s. There are no clues in Turner's *Chichester Canal* that the subject is a canal rather than an ordinary navigable river running across low ground – there are no locks, none of the bustle evident for example in *Kirkstall Lock*. Instead, Turner has portrayed it as he had portrayed the Thames *c*.1806–10, as a warm evocation of a rural sunset. Egremont had invested heavily in the construction of the Chichester Canal, which opened in 1823 as part of the final link in a chain of canals linking London and Portsmouth. The canal was a commercial failure, and Egremont withdrew from it in 1826 at great financial loss to himself.

Egremont also invested in the construction of the Chain Pier at Brighton, which opened in 1823. This by contrast was an enormous success, using new construction technology – the suspension bridge principle – to enhance one of the most fashionable seaside resorts in England. Where the only sign of life and movement in *Chichester Canal* is given by the moorhens landing on the water on the right, in *The Chain Pier, Brighton* the pier is the centre of bustling human activity – fishing craft, rowing boats and a steamboat approaching the pier with its load of passengers to land. Hung adjacent to each other as they were, on a line with Turner's two paintings of Petworth Park,[10] it may be that these were intended by Turner, and Egremont also perhaps, as a reminder of the fallibility of human endeavour. This was a subject dear to Turner's heart, on which he dwelt at length and lovingly in his poetry – 'The Fallacies of Hope', for example.[11] There may be an added and intentional irony in the choice of subjects since the project that failed

was a working commercial canal, and the one that succeeded was built for the pleasure of visitors to the seaside.[12]

In 1830 Turner travelled to the English Midlands to gather material for more watercolours for his *England and Wales* series. Besides studies in Worcester, Kenilworth and Warwick, which led directly to major watercolours for engraving, his sketchbooks also reveal small, atmospheric pencil drawings of industrial landscapes and townscapes in Birmingham (fig.102), Coventry and Dudley. These provided the raw material for Turner's evocation of the beautiful hellishness of industrial development in Britain, the watercolour *Dudley, Worcestershire* (fig. 101). The composition of this picture, with its ruddy glow to one side, parallels Turner's treatment of lurid industrial scenes from *Limekiln at Coalbrookdale* (fig. 36) to

fig. 97 *Keelman Heaving in Coals by Moonlight* 1835 oil on canvas *National Gallery of Art, Washington. Paul Widener Collection*

fig. 98 *Kirkstall Lock, on the River Aire* 1824–5 (cat. 82)

Shields, on the River Tyne (fig. 96). But however unpleasant life in Turner's *Dudley* might appear to be, and however final might be the change he evokes in the picture from an England dominated by the church to one driven by industry,[13] there is no doubt that Dudley is successful. The town has a product, iron and ironware; it has a thriving barge trade; and the factories are busily creating an economy. All this information is given in the picture and, in the first published edition of Robert Wallis's engraving it was augmented by text written by Hannibal Evans Lloyd: 'The neighbourhood abounds in mines of coal, iron-stone, and limestone, which furnishes employment for a great number of the inhabitants.'

One of the most extraordinary engineering ventures of the nineteenth century was underway in the late 1820s just across the Thames from the Ship and Bladebone, an inn Turner owned in New Gravel Lane, Wapping. This

was the Thames Tunnel, designed by Sir Marc Isambard Brunel, and overseen by his son Isambard Kingdom Brunel. Tunnelling began at Rotherhithe on the south bank of the river in December 1825, with the aim of reaching its northern entrance at Wapping, two hundred yards from the Ship and Bladebone. In the event work ceased in January 1828 after a catastrophic flood when the tunnel was halfway across, and did not resume for seven years. The tunnel finally opened to great national acclaim in 1843.

There is a curious oil painting in the Turner Bequest which has been given the title *A Vaulted Hall* (fig. 103). Although some evidence may suggest otherwise, it is worth looking at this as a possible picture of the interior of Brunel's tunnel. On the evening of 10 November 1827, a party of fifty of the great and the good sat down for a feast in one of the shafts, and one hundred and twenty of Brunel's tunnellers dined in the other to prove to the public, after a widely reported flood, that the tunnel was safe.[14] The tunnel was brilliantly lit by

fig. 99 *Chichester Canal* c.1828 (cat. 84)

fig. 100 *The Chain Pier, Brighton* c.1828 (cat. 85)

fig. 101 *Dudley, Worcestershire c.1832* (cat. 86)

gas, and the band of the Coldstream Guards played as the diners feasted. Though painted freely, there is a strong suggestion of many figures in the main arch, and more under the arch to the left. Other figures appear to pass backwards and forwards between the two, suggesting the movements of waiters. Arguing against this interpretation is the fact that the profile of the arches is not quite right (other illustrations of the tunnel indicate that its walls curved inwards towards the floor); there is no trace of the crimson draperies that Brunel hung for the occasion in the side arches;[15] and there may be a suggestion of a non-existent third arch on the right. However, if this glow is reasonably read as gas-light, the painting may indeed be properly titled *Banquet in the Thames Tunnel, 10th November 1827*.[16]

Turner and Isambard Kingdom Brunel were well known to each other, though documentary evidence of

personal contact has yet to be discovered.[17] They had many friends in common, particularly A.W. Callott and J.C. Horsley, whose sister Brunel married. Brunel commissioned some of Turner's friends in the late 1840s to paint pictures of Shakespearean subjects for the 'Shakespeare Room' in his house in Duke Street, St James's. Among these were C.R. Leslie, Clarkson Stanfield, and J.C. Horsley.[18] There are enough references in Brunel's journals to demonstrate his closeness to Royal Academy and Royal Society members, and, like Turner, Brunel was an active member of the Athenaeum Club.[19] Owning a working inn as he did so near to the anticipated Wapping entrance of the Thames Tunnel, Turner stood to benefit from the increase in trade and land-value that would accrue when the Tunnel opened,[20] and he, like any member of the public on payment of a fee, could enter the tunnel from the Rotherhithe end to see the works. A painting of the interior of this great engineering achievement at a proud moment would be an ideal subject for him. Nobody was

fig. 102 *Birmingham – Landscape Studies* 1830,
from *Kenilworth* sketchbook (cat. 97)

fig. 103 *A Vaulted Hall – perhaps 'Banquet in the Thames Tunnel, 10th November 1827'* (cat. 83)

to know that two months after the banquet, on 12 January 1828, the Tunnel would flood disastrously. If Turner's painting does represent the banquet of November 1827, we need not be surprised that he left it unfinished.

Turner's other painting with Brunel associations is *Rain, Steam, and Speed – the Great Western Railway* (fig. 104). This work pays homage not to one, but to two great technical advances, just as in *Life-Boat and Manby Apparatus* (fig. 86), where Turner rolled a quadruple homage to Captain Manby into one painting. When he decided to show his admiration, Turner did not stint it.

The Great Western Railway, built by Brunel from London to Bristol in only six years, 1836–1841, was a heroic achievement. Before work began, Brunel wrote that he was the engineer to the finest work in England – 'no idle boast', observed his biographer L. T. C. Rolt.[21] Constructing the line was fraught with difficulties: among many other obstacles, Brunel had to make a long cutting at Sonning, tunnel through Box Hill near Chippenham and bridge the Thames near Maidenhead.

The Thames Commissioners forbade Brunel to obstruct the towpath or to narrow the navigation channel beside it, but required him to use one central pier only, and place it on the island at that point of the river. Given the level of the line and the position he chose for his crossing, however, Brunel had created his

fig. 104 *Rain, Steam, and Speed – the Great Western Railway* exh. 1844 (cat. 89)

own problem and had therefore to find his own solution. To meet the height requirement, he devised a pair of elliptical arches, the largest and flattest that had ever been built in brickwork (fig. 105). Brunel's critics howled that the bridge would collapse; it did not, and the first passenger train crossed it in July 1839. There were exceptional storms the following autumn and winter, and the bridge was again widely expected to collapse, but it did not.[22] In choosing to depict a steam engine passing both along Brunel's line *and* over his Maidenhead Bridge *and* in a violent storm, Turner is allying himself directly with the engineer, and publicly applauding his triumph.

The structure of *Rain, Steam, and Speed* echoes almost precisely the composition of *Juliet and her Nurse*

(exh. 1836; private collection), down to the viewpoint being just off the edge of the major diagonal and the repetition a minor diagonal on the left. This is the old multi-arched road bridge built in the late eighteenth century by Sir Robert Taylor. Brunel's bridge appears to carry only one track, rather than the two that Brunel laid, though there is considerable ambiguity in the lines of paint. The narrowness of the bridge emphasises the speed of the train, which passes high above a ploughman trudging through the rain behind a horse-drawn plough. The ploughman, and a hare running for its life along the track, vividly register by contrast the nature of the new technology that has burst upon the world. This passes in a blur, defying the elements and leaving us with what is in effect a retinal after-image of the train rushing by.[23]

Though unfinished, Turner's *The Thames above Waterloo Bridge* (fig. 106) is clearly as populous as his pair of completed London Thames subjects, *The Burning of the*

Houses of Lords and Commons, 1834 (exh. 1835; Philadelphia Museum of Art; Cleveland Museum of Art). At 3 × 4 ft, it is the same size as they and of a favourite compositional type, paralleling the Carthage subjects, that Turner exhibited regularly. With a twin-chimneyed steamboat on the left belching smoke beyond a richly caparisoned craft, flags flying and dozens of rowing boats waiting about, something important is evidently about to happen. Butlin and Joll have suggested that *The Thames above Waterloo Bridge* may be Turner's projected answer to Constable's picture, *Waterloo Bridge from*

Whitehall Stairs, June 18th 1817, showing the opening of Waterloo Bridge and exhibited fifteen years after the event in 1832.[24] Constable had in 1831 and 1832 been the butt of Turner's teasing and anger. They were not soulmates, and Turner was of the temper to try to upstage or at least challenge Constable.[25] A date in the early 1830s is perfectly reasonable for Turner's painting on stylistic grounds, but the subject cannot be a historicising account of the opening of Waterloo Bridge, if only because there were no twin-chimneyed steamboats in 1817.

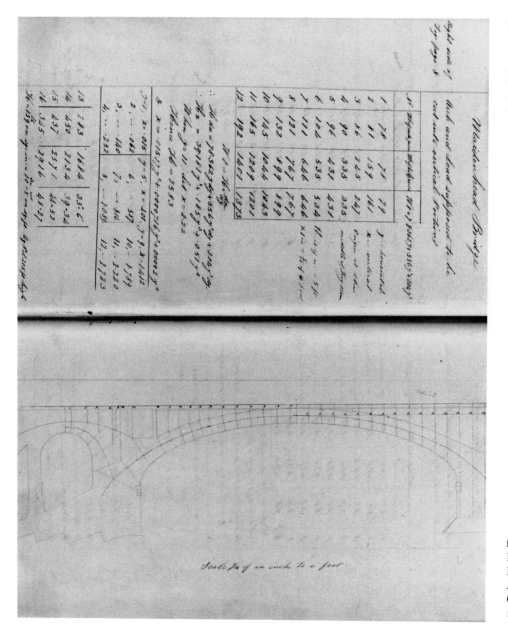

(*right*) fig. 106
The Thames above Waterloo Bridge ('Waterloo Bridge – The Procession before the Opening of London Bridge, 1st August 1831')
*c.*1832 (cat. 87)

fig. 105
Isambard Kingdom Brunel, *Arch Analyses for Maidenhead Bridge*, 1837 General Calculation Book 1834–41 (cat. 90)

The big news in bridges in 1832 was not Waterloo Bridge, but the rebuilt London Bridge. This had been opened by the King on 1 August 1831, amid pomp and celebrations, as the climax of a river procession from Somerset House (beside Waterloo Bridge) downstream to London Bridge. The *Gentleman's Magazine* reported that 'two parallel lines of vessels were formed into a passage of about 150 feet wide, consisting of a double, and in many cases a triple, line of barges, steamers, yachts, and craft of every description, which extended from the upper water-gate of Somerset House next Waterloo Bridge'.[26] Turner was present at the opening and is shown as one of the dignitaries welcoming the King and Queen in the painting George Jones RA exhibited at the 1832 Royal Academy exhibition, *The Royal Procession at the Opening of London Bridge, 1831* (exh. 1832; fig. 109).

While a challenge to Constable with fig. 106 is entirely possible, it is as likely to be a good-humoured response on Turner's part to his friend George Jones's painting of his *London Bridge* subject in 1832. That same year the pair had challenged each other to paint *Shadrach, Meshech and Abednego in the Burning Fiery Furnace*, and although only one subject is recorded in Jones's account of the jest,[27] they may have competed again, Turner this time failing to complete his attempt.[28] What Turner's painting, which we might now retitle *Waterloo Bridge – the Procession before the Opening of London Bridge, 1st August 1831*, all too plainly shows is not the clearly lit, sparkling Thames of Constable's subject, but the reality of the river in the early 1830s with smoke from steamboats and from firing cannons 'kept up without intermission along the whole line of the river'[29] on this celebration day.

In the mid-1840s Turner embarked on a series of

paintings of a subject matter that was entirely new to him. His four whaling paintings, exhibited in 1845 and 1846, were encouraged and probably prompted by his friendship during this period with the whaling entrepreneur Elhanan Bicknell (1788–1861).[30] Bicknell had made a great deal of money out of whaling, and although his company, Langton and Bicknell, did not own whalers, it flourished through developing a new process of refining spermaceti oil, highly prized as an oil for lamps and candles, and had financial interests in ships working in the Pacific and Antarctic waters.

The two whaling subjects in the Tate collection, *Hurrah! for the Whaler Erebus! another Fish!* and *Whalers (Boiling Blubber) Entangled in Flaw Ice* (both exh. 1846; figs. 111, 112) have the static composition of figures in boats around a central void, which is a chilly variation on Turner's *Waterloo Bridge* and his Cowes and Carthage paintings. There are immediate resemblances in these scenes of industry at southern latitudes to *Shields* (fig. 96) and *Keelmen Heaving in Coals by Night*. Robert K. Wallace has given a clear demonstration of the Antarctic sources for the whaling pictures and has revealed also some of the ways in which Turner was primed with the technical information he needed to paint the subjects.[31]

These come down to contact with a small number of individuals, central among whom was Elhanan Bicknell, in whose house at Herne Hill Turner was a regular and

(*above and left*) figs. 107, 108 *London Bridge under Construction* c.1827, from *Isle of Wight* sketchbook (cat. 96)

welcome guest in the 1840s.[32] Conversations with Bicknell must as a matter of course have encompassed whales and whaling, subjects which, as I have suggested elsewhere, were a rational evolution from the monster imagery of Turner's young manhood and from the writhing serpent allusions of his early poems.[33] Wallace also points out that among Turner's fellow members of the Athenaeum Club were the sailor James Clark Ross, Captain of *HMS Erebus* on its voyage with *HMS Terror* to Antarctica from 1839 to 1843, and the zoologist J.E. Gray, who co-edited an account of the *Zoology of the Voyage of H.M.S. Erebus and Terror*, published from 1844.

Turner owned a copy of 'Fishes', part 5 of the *Zoology* which appeared in 1845, and might also have known about Gray's subsequent volume in the series, 'On the Cetaceous Animals' (i.e. whales etc.), published in 1846.

Wallace also reveals the link between Turner and Joseph Dalton Hooker, who sailed as assistant surgeon on the *Erebus* and made many drawings on the voyage, some of which were engraved for publication in Captain Ross's official account.[34] Hooker, who later became a distinguished botanist, was the grandson of Turner's good friend the Yarmouth banker, amateur botanist and prodigious correspondent, Dawson Turner. Dawson

fig. 109 George Jones,
*The Royal Procession
at the Opening of London
Bridge, 1831*
1832 oil on canvas
*Sir John Soane's Museum,
London*

Turner and J.M.W. were in correspondence during the early 1840s, at the same time as Hooker was writing to his grandfather with vivid descriptions of the voyage. In a letter written from the Athenaeum on 29 November 1842 J.M.W. reveals with a flourish of exasperation to Dawson Turner how his mind is taken over by new ideas and how he is driven to express them. 'You ask me what "are you doing",' J.M.W. writes, and answers the question: 'endeavouring to please myself in my own way if I can for all my determination to be quiet some fresh follerey comes across me and I begin what most probably never to be finish'd.'[35] The 'fresh follerey' that came across Turner in the late autumn of 1842 may have been connected to the paintings he exhibited in the 1843 Academy – his 4 × 6 ft mahogany panel *The Opening of the Wallhalla, 1842*[36] or his pair of experiments in shade and darkness, and light and colour, *The Evening of the Deluge* and *The Morning after the Deluge*.[37] Whatever the case,

there is a certain amount of resigned irony in the letter which gives yet more evidence for what we already know of Turner's insatiable interest in the world around him and in new ideas and imagery which drove him on to new means of expression. Whaling among the Antarctic ice, such an exotic combination of distance, cold, danger and human endurance, was as potent a source of imagery for him as a limekiln at Coalbrookdale or the Great Western Railway.

There is, however, a further twist to the tight relationship of painter, scientist and whale that short-circuits all those outlined above. By the mid-1840s Turner and the zoologist Richard Owen (1804–92) were already good friends. They had met probably in the late 1830s at a dinner given by W.J. Broderip (1789–1859), a lawyer, amateur naturalist, and active collector of Turner, and over the years Turner made several visits to Owen and his wife in their apartments at the Royal

fig. 110 *Freiburg with Suspension Bridge* 1842, from dismembered *Freiburg* sketchbook (cat. 88)

fig. 111 *'Hurrah! for the Whaler Erebus! Another Fish!'* exh. 1846 (cat. 93)

College of Surgeons.[38] There is a specific reference to one visit in August 1845 when Mrs Owen recalled she translated part of the programme of the 1845 Munich Exhibition for Turner, 'as he is thinking of sending them a picture'.[39]

Owen was becoming a leading figure in the movement to establish public collections of zoological material, an ambition that kept him at the forefront of natural history politics, and the driving force behind the creation of the Natural History Museum in South Kensington. Writing in 1839 to J.W. Lubbock, Treasurer to the Royal Society, Owen urged the Society to recommend that the whaling industry be involved in collecting specimens for scientific study:

The museums of this country would be greatly enriched and Natural History advanced by any plan which would give encouragement to … the Sperm Whalers to collect the rare objects that they may meet with, and to keep records of the Natural Phenomenas, habits and peculiarities of the living animals which they may observe. Perhaps the simplest plan of encouragement would be to propose premiums for certain rare animals, and for journals or observations, and to encourage the cooperation of the South Sea- and Whaling Merchants in the distribution of printed lists of such premiums with Instructions for Collecting among their Officers and Crews.[40]

As a whaling entrepreneur, Elhanan Bicknell will have been well aware of the development of Owen's proposal. A few weeks after Owen had written to Lubbock, the Royal Society published their *Report of the President and Council … on the Instructions to be Prepared for the Scientific Expedition to the Antarctic Regions*, which urged that

fig. 112 *Whalers (Boiling Blubber) Entangled in Flaw Ice, Endeavouring to Extricate Themselves* exh. 1846 (cat. 94)

specimens of marine invertebrates, fishes, reptiles, birds and mammalia be collected. Owen's intervention was evidently timely and effective, and although Turner may have bumped into Captain Ross and J. E. Gray at the Athenaeum, and discussed young Joseph Hooker's letters with his grandfather, the most certain link between Turner and his choice of whaling subjects at this particular period of his career is his friendship both with the driven zoologist Richard Owen and the hospitable whaling entrepreneur Elhanan Bicknell.

Visiting Owen as he evidently did on a number of occasions, Turner had every opportunity to discuss the widest areas of natural history, and this must have been among the purposes of their meetings. Certainly Turner showed a special regard for Owen at what may have been only their second meeting, for, having invited him to his studio to see his paintings a few days after their first dinner together, Turner did something

which Owen declares that none of his artist friends would ever believe. Turner offered him a glass of wine! It was while they were coming downstairs that he first observed symptoms of a struggle going on in Turner's bosom. When they were passing a little cupboard on the landing this struggle reached a climax. Finally, Turner said, 'Will you – will you have a glass of wine?' This offer having been accepted, after a good deal of groping in the cupboard a decanter was produced, of which the original glass stopper had been replaced by a cork, with the remains of some sherry at the bottom. This Owen duly consumed, and shortly afterwards took his leave, with many expressions of the pleasure that the visit had afforded him and a disturbing conviction that the sherry might lurk indefinitely in his system.[41]

Turner's final and most dramatic evocation of industry combines qualities that had run indelibly through his life's work like the veins in marble. Allegory and bare fact, history, heroism, and patriotism are all rolled together in *The Hero of a Hundred Fights*, exhibited at the Academy in 1847 (fig. 115). We have already discussed the first incarnation of this canvas, as a dark interior of a forge with a woman seated beside a big wheel, painted by Turner towards the end of the first decade of the century. Suffice it to say here that by 1847 it was unsold and probably sitting among the stacks of pictures at the back of Turner's Gallery. We do not know why Turner had nothing ready for the 1847 Academy, but his strength of mind and rate of production was usually such that this is likely to be an indication that he had been seriously unwell over the winter of 1846/47. He had a chronic condition that affected him for months, and also led him to show nothing in the 1848 exhibition.[42]

Turner probably reworked the painting during the Varnishing Days at the Academy, following his regular practice. Under his hand the subject transformed from the dull forge scene to a blazing allegory of the casting of the equestrian statue of the Duke of Wellington by Matthew Cotes Wyatt.[43] For many months in 1846 the papers had been full of news of the progress of the statue, which had been in production at Wyatt's studio in Dudley Grove House, Harrow Road, since it was commissioned in 1838 by the Treasury Board as the Wellington Military Memorial.[44] It was the subject of wide public controversy, loved and loathed in equal measure, supported by *The Times* and the *Illustrated London News* and condemned by the *Athenaeum* and the *Art Union*.[45] The *Illustrated London News* had run a number of articles illustrated with vivid wood-engravings of work in progress (fig. 116). These culminated in a long report published on 3 October 1846 of the statue's transportation to Hyde Park Corner, accompanied by up to one thousand soldiers and four military bands playing 'See

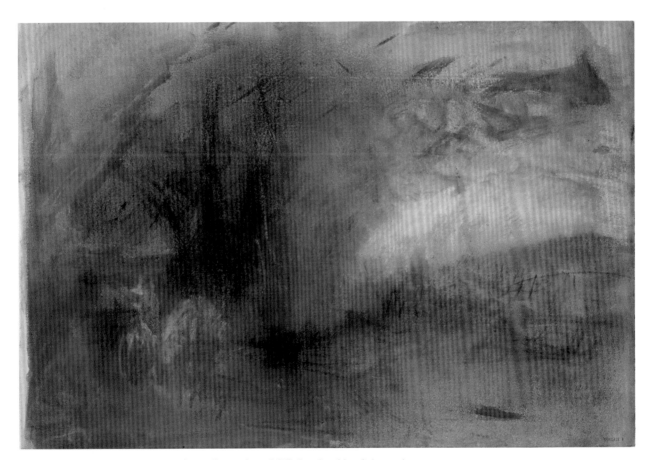

fig. 113 *Burning Blubber* mid 1840s, from dismembered *Whalers* sketchbook (cat. 91)

fig. 114 *Burning Blubber* mid 1840s, from dismembered *Whalers* sketchbook (cat. 92)

the Conquering Hero Comes', and its raising to the top of Decimus Burton's arch.[46] The installation of the statue was a national event, witnessed by a host of the nobility and senior politicians, including Prince George, royal dukes and the Prime Minister Lord John Russell. The Queen watched the installation from Apsley House. Wyatt's statue was, in the words of the *Illustrated London News*, 'destined … for centuries to commemorate the bravery of the British Hero; the skill of the British Artist; and the gratitude of the British Nation.'[47]

Turner's evocation of the opening of the furnace to reveal the completed cast is an invention for dramatic effect. It was not cast whole – sculpture of that size cannot possibly be – but in pieces, and later assembled. Further, despite the evidence of the painting, Wyatt's furnace was in a pit, not at floor level. It is possible that Turner was one of the 'considerable party of scientific

and literary men and artists' who witnessed the pouring of seventeen tons of molten metal into the mould of the forequarters of the horse and so had first-hand knowledge of the event. The *Literary Gazette* described the scene:

The flow of so large a quantity of molten metal from the furnace to the receptacle whence it descends to fill the mould is a very grand and remarkable phenomenon, affording a perfect idea of a volcanic eruption. The furnace … is tapped in the interior of the building by a long iron rod being beat against a lower vent; and the imprisoned fluid gushes out with tremendous fury and wonderful beauty into a channel prepared for its conduct. The dazzling red stream throws up clouds of vapour of every prismatic hue, the green tinge prevailing; but blues, yellows, and various gradations of red, rolling along both in these clouds and in flames emitted from, accompanying and hovering over, the lava torrent.'[48]

In a note to the Academy catalogue entry for *The Hero of a Hundred Fights* Turner wrote: 'An idea suggested by the German invocation upon casting the bell: in England called tapping the furnace.' John McCoubrey

fig. 115 *The Hero of a Hundred Fights* c.1806–7, reworked and exhibited 1847 (cat. 95)

has pointed out that Turner's 'invocation' was probably the epigraph 'Vivos voco. Mortuos plango. Fulgara frango' to Schiller's *Song of the Bell*, which was already well known in Britain in the early 1840s.[49] Translated, the epigraph reads 'I call to the living. I mourn for the dead. I strike thunderbolts from the sky,' three phrases that could with justification be applied to Wellington's career as a soldier. There is a clear parallel in the figure of the mounted duke with the central figure of *The Angel Standing in the Sun* (Tate Gallery) which Turner had exhibited at the Academy the year before, with lines from the Book of Revelation:

And I saw an angel standing in the sun; and he cried with a loud voice, saying to all the fowls that fly in the midst of heaven, Come and gather yourselves together unto the supper of the great God; That ye may eat the flesh of kings and the flesh of mighty men, and the flesh of horses, and of them that sit on them, both free and bond, both small and great.'[50]

It is remarkable that the year after exhibiting *The Angel Standing in the Sun* Turner should exhibit a painting of a mighty man seated on a horse consumed by fire. Turner was no preacher, and if there is any connection between these two pictures it can only be visual. The powerful circular form in which the statue of the duke appears, does, however, recall some words of Sir Humphry Davy in a passage we have looked at elsewhere:

in a limited sphere of vision, in a kind of red hazy light similar to that which first broke in upon me in the Colosseum, I saw moving round me globes which appeared composed of different kinds of flame and of different colours. In some of these globes I recognised figures which

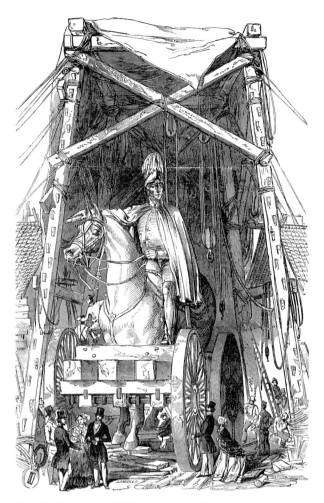

fig. 116 'The Great Wellington Statue: The Statue at Mr Wyatt's Foundry', *Illustrated London News*, 3 October 1846

put me in mind of the human countenance, but the resemblance was so awful and unnatural that I endeavoured to withdraw my view from them.'[51]

There is not enough solid evidence to show Turner knew of Davy's writing, but as we have seen in Chapter 4 Davy is a possible source for Turner's predilection for bubbles in his pictures. Davy was a great chemist, to whom dramatic, fiery experiments were stock-in-trade. To use Davian imagery, both written and practical, in the painting of the firing of the statue of the Duke of Wellington, alongside an allusion to Schiller's 'I call to the living. I mourn for the dead. I strike thunderbolts from the sky,' is potent art indeed. In combining these ideas with the remains of a painting of old technology, made perhaps forty years earlier, Turner is expressing how the Duke of Wellington, born like Turner himself in the era of water-powered mills and foundries, had come to glory in the new Victorian age as *The Hero of a Hundred Fights*.

7

THE LIVING EARTH: GEOLOGY AND MAGNETISM

In the true spirit of Romanticism, the mountains Turner painted in the early decades of his career could be breathtaking beyond reason. *The Pass of St Gothard* (fig. 117), painted soon after Turner's 1802 tour of France and Switzerland, has the viewer standing, with Turner, in the middle of the Devil's Bridge above the vertiginous drop to the valley floor.[1] The clouds are below. Combining high emotional drama with scientific truth, Turner's treatment of the rockface is as geologically accurate and descriptive a piece of painting of schist as one could hope for. In watercolours of Swiss glacier subjects painted after the same trip Turner's treatment is such as to make a scientist swoon. The botanist Charles Lyell (1767–1849), father and namesake of the pioneer geologist, wrote to Dawson Turner in 1815: 'When we meet in Town I can take you … to Walter Fawkes Esq Upper Harley Street who has the most suberb *drawings* of the Glaciers by Mr Turner RA that I ever beheld.'[2]

Nearly thirty years later, William Buckland (1784–1856), Dean of Westminster, a Fellow of the Royal Society and a former President of the Geological Society, urged Turner to consider as a subject the raising of the stores and treasure lost when HMS *Thetis* sank at Cape Rio, Brazil, in 1830. He lent Turner his copy of Thomas Dickinson's *Narrative of the Operations for the Recovery of the … Treasure* (1836) to tempt him, but, alas, as Turner remarked in his reply, there were 'pictorial misfortunes' – the ship was at the bottom of the sea, and there was no treasure to be seen.[3] Buckland had misread the direction of Turner's lifelong priorities as an artist, and had hoped the drama of the geological setting alone would drive him to paint the subject. This, however, left little to Turner's interpretative imagination:

The strong glare cast from the torches on every projection of the stupendous cliffs, rendered the deep shade of their indentations and fissures more conspicuous – their darkness more visible, and notwithstanding there was a flickering brightness thrown upon some parts, a solemn gloom pervaded the whole, which was much heightened by contrast with the brilliant whiteness of the foam on the rocks beneath. The rushing of the roaring sea into the deep chasms, produced a succession of reports like those of cannon, which were multiplied by echoes from the surrounding cliffs.[4]

Although he was refused this time, the fact of Buckland's approach to Turner, and the earlier instinctive response of Lyell to the glacier subjects, are clear indications of the professional respect that these men had for the fêted artist, and of their appreciation of his abilities as a channel for the spreading of scientific knowledge.

Although this is the earliest-known contact between Turner and Buckland, the two had many friends in common and must have known one another for years. Buckland had been Professor of Mineralogy at Oxford since 1813, where he was a popular, eloquent and witty lecturer, and a canon of Christ Church from 1825. Turner's own fondness for Oxford (and indeed Christ Church) and his regular visits there are well documented, so this is but one of their possible meeting places. Buckland was a close friend of Chantrey[5] and of Turner's patron and stockbroker Charles Stokes, of whom more later, and in the early 1830s was greatly admired by the young John Ruskin, whose early scientific leanings took him towards the study of geology.

Another of Turner's geologist friends was John MacCulloch (1773–1835), who according to Thornbury met Turner at the home of the watercolour painter W.F. Wells (1762–1836).[6] This will have been about 1814, when Wells and MacCulloch were together on the staff of the East India Company's Military College at Addiscombe in Surrey. As we shall see, it is quite possible that they met up to four years earlier. Wells's daughter Clara, whom Turner had known since her childhood,[7] was herself aware that Turner was 'greatly interested in the science of geology' and remarked that MacCulloch was 'delighted with [Turner's] acute mind',

fig. 117 *The Pass of St Gothard c.*1803–4 oil on canvas *City Museums and Art Gallery, Birmingham*

fig. 118 Robert Wallis, after J.M.W. Turner, *Stone Henge,
Wiltshire* 1829, from *Picturesque Views in England and Wales*
(cat. 101)

observing that he 'would have been great in any- and
everything he chose to take up. He has such a clear,
intelligent, piercing intellect.'[8]

Charles Stokes (1785–1853) was a distinguished
amateur geologist and determined collector. Luke
Herrmann has observed that Stokes deserves to be
ranked as one of Turner's most valued friends and
patrons, almost on a par with Walter Fawkes and
Lord Egremont.[9] Turner, Stokes and their mutual
friends Francis Chantrey and George Jones were part
of the same informal social circle who met in private
gatherings, or at the Royal Academy, the Royal Society
or the Athenaeum Club. George Jones recalled that
Stokes was 'a constant guest' at Chantrey's Sunday
gatherings, where 'it was a common occurance to meet

men distinguished by science or literature … In the
evening, the specimens of [Stokes's] minerals and fossils
were examined and the instructive allurements of the
microscope filled every moment with gratification.'[10] In
addition to his geological collection, Stokes had a fine
collection of paintings and drawings, including many
watercolours by Turner, and became one of the first
serious collectors of drawings and proofs from the *Liber
Studiorum*. His deliberate and scientific approach to tasks
in hand led him to make the first catalogue of the *Liber*
images and to draft an interpretation of Turner's
opaque personal system of marking his proofs.[11] Among
the Turners in Stokes's collection was a considerable
number of subjects with a geological slant, such as *The
Mew Stone* (National Gallery of Ireland), *Minehead* (Lady
Lever Art Gallery, Port Sunlight) and *Bridport, Dorsetshire*
(Bury Art Gallery). Another is the line engraving *Stone
Henge* (no. 118), a romantic view of the stones silhouetted
against a thunder-filled sky, yet further indication of the

fig. 119 W.B. Cooke, after J.M.W. Turner, *Lulworth Cove, Dorsetshire* 1814 from *Picturesque Views on the Southern Coast of England* (cat. 99)

pleasure that Turner's understanding of geological form gave to professional and amateur geologists alike.

Turner's poetry reveals yet more evidence of the keenness of his eye for the structure of the earth and, significantly, his knowledge of technical terms. His extended poem in the *Devonshire Coast No. 1* sketchbook includes the lines:

A little hollow excavated round
To a few fishing boats give anchorage ground
Guarded with bristling rocks whose strata rise
Like vitrified scoria to southern skies
Called Lulworth cove ...[12]

Elsewhere in the 'Devonshire' verses he uses phrases such as 'Horizontal strata, deep with fissure gored',[13] 'bristled front Basaltic like or ranged / Rock piled on rock in many forms arranged'[14] and 'beach of stone / Of

Granite Marble Slate and Lime stone formd',[15] all of which indicate an understanding of geological language derived, at the very least, from informal reading, from geologists themselves and possibly also from attendance at geology lectures.[16] These verses were written in 1811, the year the Geological Society published its first volume of *Transactions*, which included J.F. Berger's important articles *Observations on the Physical Structure of Devon and Cornwall* and on *Hampshire and Dorsetshire*.[17] The emphasis on geology in Turner's verses suggests that the Society's new activity, and his geologist friends, had a specific impact on the rapid development of his interest in the subject, and may even have directed it at this time.

John Gage has suggested that it was probably MacCulloch, one of the original Council Members of the Geological Society at its foundation in 1807, who gave Turner the first two volumes of the Society's *Transactions* (1811–14) which were found in Turner's library at his death.[18] This may be so, but Turner's interest in the subject was already so intense that we

fig. 120 *Geological Formations at Lulworth Cove* 1811, from *Corfe to Dartmouth* sketchbook (cat. 105)

need not assume that he waited for a gift of the books. The volumes include eleven articles by MacCulloch himself, one at least of which, *On Staffa*, may have helped to prepare the ground for Turner's interest in this Scottish island which culminated in his painting *Staffa – Fingal's Cave* (exh. 1832). See Chapter 5, above.

The engraving *Lulworth Cove, Dorsetshire* (fig. 119), with its purposeful delineation of the structure of rock strata and cliffs, is a particular example of a quality that stands out as characteristic of the *Southern Coast* images which Turner embarked upon in 1811. They mark the maturity of a lifelong engagement with mountain, hill and rock forms that is underpinned by a certain knowledge of the type of rock forms depicted, their difference from any other, and the dynamics of the folding of the earth. Turner's friendships with geologists must be a central factor in this, as must the coincidental publication of Berger's two lengthy and specific articles on the geology of south-west England. In his clear understanding of perspective and Euclideian geometry we have seen that Turner could be considered an accomplished geometer (Chapter 2); we might now add to this that through his unique means of expression Turner was also a geologist. Seven years later he demonstrated in his lectures at the Royal Academy an awareness of the synthesis between geometry, geology and the insect

(left) fig. 121 *Close-Up View of Rock Face* 1811, from *Stonehenge* sketchbook (cat. 106)

(right) fig. 122 *Interior of Fingal's Cave, off the Isle of Staffa* 1831, from *Staffa* sketchbook (cat. 107)

world: 'the cell of the Bee and the Basaltic mass display the like Geometric form, of whose elementary principles all Nature partakes.'[19]

Watercolours from the early 1830s which Turner painted specifically for engraving, such as *Glencoe* (fig. 123; engraving, fig. 124), *Loch Coriskin* (National Galleries of Scotland; engraving, fig. 125b) and *Fingal's Cave* (private collection, USA; engraving, fig. 125a) show with what passion his theories of geometric form met his graphic descriptions of rocks, mountains and storms. Figs. 125(a) and (b) were made to illustrate the 1834 edition of Sir Walter Scott's *Poetical Works*, the images being conceived during and shortly after Turner's 1831

visit to Scotland and the Western Isles. Here he renewed his professional relationship with Scott, and travelled with him briefly around the Lowlands.[20] In Turner's *Loch Coriskin* (correctly, Coruisk) the form of the mountains rhyme with the coming storm, powerfully evoking the lines in 'Lord of the Isles', in which Scott's hero first comes upon the loch:

Seems that primeval earthquake's sway
Hath rent a strange and shatter'd way
Through the rude bosom of the hill,
And that each naked precipice,
Sable ravine, and dark abyss,
Tells of the outrage still.[21]

In Turner's *Fingal's Cave* the arch creates a vortex formed by countless pieces of carefully observed basalt, which may stand as a graphic metaphor for the manner in which Scott's writings, Turner's illustrations and the geological researches of John MacCulloch interlink. MacCulloch and Scott were old friends. MacCulloch's analysis of the structure of the landscape in *Highlands and Western Isles of Scotland* (1824) evolved out of a correspondence with Scott, one of a number Scott had with scientists.[22] In these letters MacCulloch wrote a clear descriptive account of the structure of Fingal's cave, and the first prose description of Loch Coriskin on the Isle of Skye in which he 'felt like an insect amidst the gigantic scenery, and the whole magnitude of the place became at once sensible'.[23] Turner's friendship with MacCulloch has already been discussed, and although his relationship with Scott was not a close one, the two had willingly bound themselves together in literary and illustrative projects since 1818.

Another article in Turner's volumes of the Geological Society *Transactions* is *An Account of 'The Sulphur' or 'Souffrière' on the Island of Montserrat* by Nicholas Nugent. This may have been one prompt for Turner's painting *The Eruption of the Souffrier Mountains* (fig. 126), though according to Turner's full title the inspiration for it was a sketch by Hugh P. Keane. He, however, disappeared abroad, together with his sketch and the proofs and copper plate of the engraving, and was never heard of again.[24] Never having yet seen a volcano in eruption, Turner's approach to the subject is, if anything, uncharacteristically restrained. This is not so much an eruption, more a mountain on fire, as the title of Charles Turner's engraving, *The Burning Mountain*, explicitly puts it. Though debris shoots from the crater in *Souffrier*, this is more comparable to the firework displays painted by Joseph Wright of Derby forty years earlier than to the percipient accounts of eruptions by contemporary vulcanologists. Turner's approach to the effects of firelight on the mountain is akin to his manner of painting a horizon with the sun rising or setting behind it – the silhouette being bitten into and partially dissolved by the light.[25] For additional dramatic effect Turner has flicked loose paint at the canvas, resulting in splats and trails in the sky on the left. When Turner did see a volcano, Vesuvius, on his brief visit to Naples in November 1819, it was smoking rather than erupting, in

fig. 123 *Glencoe c.1832–3* watercolour *Museum of Art, Rhode Island School of Design. Gift of Mrs Gustav Radeke*

fig. 125 (a) Edward Goodall after J.M.W. Turner, *Fingal's Cave* 1834 (title page from Sir Walter Scott's *Poetical Works*, vol. x) (cat. 108a);
(b) Henry Le Keux, after J.M.W. Turner, *Loch Coriskin* 1834 (facing title page) (cat. 108b)

(*left*) fig. 124 *Glencoe* c.1834–6 (cat. 103)

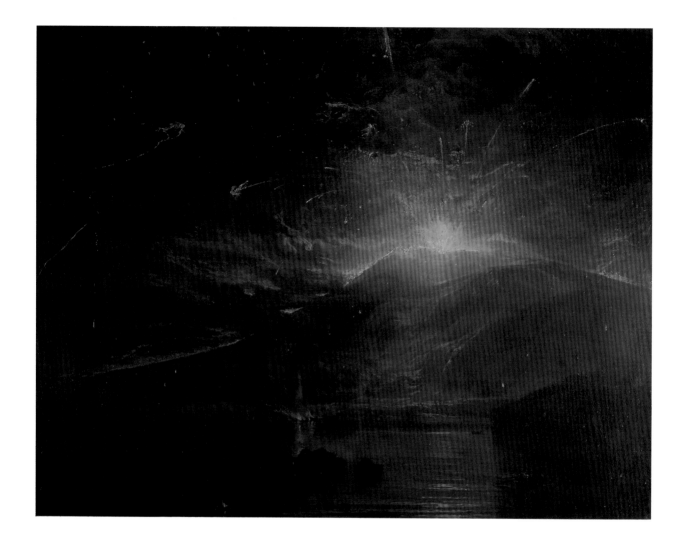

fig. 126 *Eruption of the Souffrier Mountains, in the Island of St Vincent, at Midnight, on the 30th April, 1812 from a Sketch Taken at the Time by Hugh P. Keane, Esq.* 1815 (cat. 100)

much the same condition as Mary Somerville described in a letter home in 1818. After looking down into the crater, she wrote:

My feelings were so strong of astonishment and still more of terror that I stood motionless without uttering a word. The extent is enormous, the depth unknown, the inside rough shaggy horrible, smoking fiercely from every crevice, brilliant with all colours, deep black bright red green yellow orange which all formed by the vapour, the heat and the smoke was distressing but we were determined to go round it which took us an hour, at one place we stuffed handkerchiefs into our mouths and noses and ran as fast as we could over a place that was so thick of vapour we could hardly see one another. The red flame issued from many places and the heat so great that I several times thought my petticoats were on fire.[26]

Intrepid traveller as he was, Turner would hardly have missed the opportunity of climbing up to the crater of Vesuvius, unless, in the few days he was there, it was considered too dangerous to do so. There are no surviving studies of the crater, so we might conclude that when Turner was in the area Vesuvius was grumbling alarmingly. In all its appearances in his 1819 drawings, Turner's Vesuvius is a relatively distant presence, sometimes smoking, never dominating the landscape, dramatic by implied power rather than by design. It is significant that his one set-piece composition of *Vesuvius in Eruption* (1817; Paul Mellon Collection, Yale Center for British Art) is an over-stated and recherché attempt made from another man's sketches two years before he

fig. 127 *Sketch for 'Ulysses Deriding Polyphemus'* ?1828 (cat. 102)

saw the mountain for himself.[27] Turner did, however, take one thing away from his visit to Vesuvius in 1819, a snuff box made from its lava, which he used as a palette on the tour.[28]

Turner takes a more poetic excursion into painting volcanoes in *Ulysses Deriding Polyphemus* (exh. 1829; National Gallery, London – oil sketch, fig. 127). The mountain on which Polyphemus lies is Turner's vision of Etna (named from the Greek for 'I burn'), the active volcano on the coast of Sicily. Though the fire is small in scale, the imagery is geologically possible, as Etna has historically had fissures down its sides, erupting on occasions at sea level. It is unlikely that Turner ever saw Etna, but though tentative in fig. 127 and bolder in the exhibited painting, the fires on the surface of the sea may be a conscious reference to contemporary geological thinking.

Turner will certainly have known of many of the controversies among geologists concerning the earth's formation. These included the debate between the Neptunists and the Plutonists, in which the former believed that rocks had been precipitated from a universal ocean, while the latter, led by the Scottish geologist James Hutton (1726–97), maintained that the earth was a dynamic body that functions as a heat machine. Thus granite was seen as an intrusive igneous rock rather than a crystallised sediment as the Neptunists believed. By the end of the eighteenth century the Plutonists had won the argument. One of the most potent controversies of the 1820s was the theory of the Universal Deluge, shown as proven by Buckland following his discovery of fossils and hyena bones in a cave in Yorkshire. These, he believed, were evidence of life on earth before Noah's flood.[29] In some of his cataclysmic geological subjects, such as *Loch Coriskin*, Turner seems to skirmish with graphic representations of the idea of the Deluge. Turner's reading will also have encompassed the poetry of Erasmus Darwin, whose *Temple of Nature* (1803) is a source of the idea touched upon in *Ulysses Deriding Polyphemus* that volcanoes were originally created by the rush of sea-water into the burning caverns of the Earth:

Till central fires with unextinguished sway
Raised the primeval islands into day.[30]

Turner was not alone among his artist colleagues in his fascination with geology. The sculptor Francis Chantrey, to whom marble was stock-in-trade, quite naturally had a professional interest in the subject. He had a fine geological collection, which ebbed and flowed with gifts to and from his scientist friends. Faraday asked Charles Babbage to tell Chantrey that he, Faraday, had found a piece of meteoric iron for his collection,[31] and through Stokes Chantrey arranged to pass on to Faraday a length of silicified wood weighing about half a ton.[32] C. R. Cockerell wrote in his diary, perhaps disparagingly, that Chantrey 'smattered on geology',[33] and often in his conversation Chantrey would quote

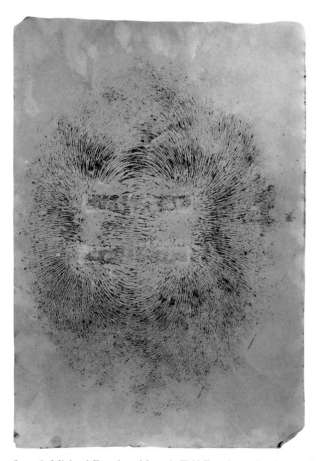

fig. 128 Michael Faraday, *Magnetic Field Experiment* 1851 (cat. 109)

Davy and the geologist William Wollaston.[34] It must have been at Chantrey's prompting that Humphry Davy noted that one of his purposes in Rome in 1819, as previously quoted, was to 'determine if the spots in Carrara marble are *iron oxide* & if they can be *obliterated*'.[35]

In an observation that is unknowingly appropriate, Benjamin Robert Haydon described Turner in 1829 as 'the Pole Star of Art, the needle where there is no variation'.[36] Research into terrestrial magnetism was at that time being greatly encouraged by the British government, the Royal Society and the army and navy. Successive voyages, under the commands of David Buchan, James Clark Ross and others, had taken place in the 1820s and 1830s to the Arctic and Antarctic to discover *inter alia* the nature, intensity and origin of the Earth's magnetic field. For the crew of one voyage, that of Sir Edward Parry in 1821–3, Mary Somerville made pots of marmalade, and found, on the ship's return, that Parry had named an Arctic island after her.[37]

Magnetism as a phenomenon attracted Turner greatly. He took particular interest in Mary Somerville's experiment to discover the magnetising power of violet light in the spectrum, read to the Royal Society on 2nd February 1826, and discussed it with the photographer J. J. E. Mayall in the late 1840s (see Chapter 1). This was long after the author had herself withdrawn the idea as mistaken, and had destroyed as many of the published abstracts as she could.[38] During the same period as Somerville carried on her magnetic researches, her friend Michael Faraday was investigating the relationship between electricity and magnetism. This culminated in his discovery of electro-magnetic induction, which he introduced to the Royal Society in a paper on 24 November 1831.[39] Lines of magnetic force can be revealed by anybody equipped with iron filings and a magnet,[40] and during later experiments into magnetic field theory Faraday made a series of sheets showing the patterns made by iron filings when placed in a variety of configurations of magnetic fields (fig. 128). Faraday conducted his experiments in his laboratory in the basement of the Royal Institution in Albemarle Street where Harriet Moore (1796/7–1884), a mutual friend of his and Turner's, painted a group of observant watercolours of the scientist at work surrounded by his equipment (fig. 9). Harriet Moore, who was a daughter of Turner's friends Mr and Mrs James Carrick Moore

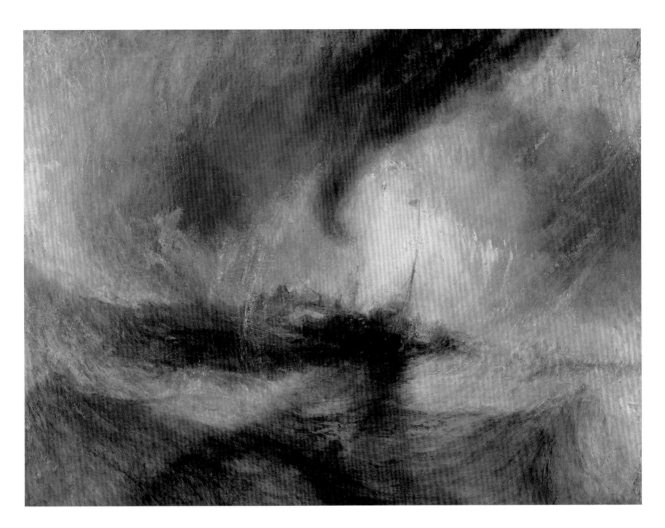

fig. 129 *Snow Storm – Steam-boat off a Harbour's Mouth Making Signals in Shallow Water, and Going by the Lead. The Author was in this Storm on the Night the Ariel left Harwich* 1842 (cat. 104)

and a regular hostess and correspondent with Turner,[41] was a staunch supporter and Member of the Royal Institution, giving money on various occasions to its Fund for Promoting Experimental Researches.[42] Harriet Moore's portrait can be seen in the audience in Alexander Blaikley's lithograph of Faraday's Christmas Lecture in 1855. Moore's skill as a watercolourist was recognised by the physicist John Tyndall (1820–93) for whom she painted large-scale lecture diagrams.[43]

Turner's regular friendship with Harriet Moore clearly made more of the direct connection between him and Faraday and another mutual friend the lithographer Charles Hullmandel (1789–1850),[44] who was also closely involved professionally with Buckland and Stokes

and described lithography as 'an art entirely founded on Chemistry'.[45] These tightly knit social circles, small but powerful magnetic fields in themselves, brought artists, writers, scientists and patrons together in a highly productive but usually unchartable flow. Magnetism was in the air in the early 1830s, in talk, experiment and writing. Looking out across the sea at St Leonard's in Sussex in 1831, the poet Thomas Campbell saw steamships and thought about magnetism:

Earth has her gorgeous towns; the earth-encircling sea
Has spires and mansions more amusive still –
Men's volant homes that measure liquid space
On wheel or wing …
Moored as they cast the shadows of their masts
In long array, or hither flit and yond
Mysteriously with slow and crossing lights,
Like spirits on the darkness of the deep.
There is a magnet-like attraction in
These waters to the imaginative power

That links the viewless with the visible,
And pictures things unseen.[46]

Looking at the same time at the same sea sixty miles around the Kent coast, Turner saw the same 'slow and crossing lights' of the steamships off Margate. During the 1830s, as we have already seen, Turner was a regular visitor to the home of William and Mary Somerville, and Mary Somerville visited his studio frequently. Magnetism, which greatly preoccupied Somerville at this time, must have been a central topic of conversation. During the early 1830s Somerville was working on *The Connexion of the Physical Sciences*. In this she looks generally at the phenomena of electricity and magnetism and observes:

Since the motion, not only of metals but even of fluids, when under the influence of powerful magnets, evolves electricity, it is probable that the gulf stream may exert a sensible influence upon the forms of the lines of magnetic variation, in consequence of electric currents moving across it, by the electro-magnetic induction of the earth. Even a ship passing over the surface of the water, in northern or southern latitudes, ought to have electric currents running directly across the path of her motion. Mr Faraday observes that such is the facility with which electricity is evolved by the earth's magnetism, that scarcely any piece of metal can be moved in contact with others without a development of it, and that consequently, among the arrangements of steam engines and metallic machinery, curious electro-magnetic combinations probably exist, which have never yet been noticed.[46]

Taking Mary Somerville's graphic idea a stage further, the Astronomer Royal George Biddle Airy (1801–92) gave a paper to the Royal Society in 1839 on experiments to correct compass deviation in iron-built ships.[47] Airy's fundamental supposition was that 'every particle of iron [in the ship] was converted into a magnet', and he ended his paper with the observation that the most remarkable result of the experiments was 'the great intensity of the permanent magnetism of the malleable iron of which the ship is composed'. The reality of magnetic forces at sea was already an accepted principle in the 1830s, part of scientists' general understanding, which included the idea that 'magnetism existing in the earth independently of it,

electric currents may be produced, not only by the earth's rotation, but by the motion of the waters on its surface.'[48]

There is already enough weight of mystery surrounding Turner's *Snow Storm – Steam-Boat off a Harbour's Mouth* (fig. 128), so it may perhaps support a little bit more. Turner claims in the title of the picture to have been in the storm on the night the *Ariel* left Harwich. This and another comment have been read to suggest that the artist was somehow lashed to the mast of the ship, Ulysses-like, to witness the storm. He was not.[49] He had witnessed storms like this around the coast many times in his life, and for him to be 'in' a storm does not mean he has to be on board a ship. It is my contention that, whatever else he may have meant by '*the Ariel*' and 'Harwich',[50] Turner is addressing a very much wider and more fundamental question, and that this is clouded if we spend too much effort in a fruitless search for an identity for the ship.

In *Snow Storm – Steam-Boat off a Harbour's Mouth* Turner is giving graphic expression to the very real lines of force that his scientist friends had showed were being emitted from all points of the earth's surface at all times. With certain knowledge of Mary Somerville's words, 'even a ship passing over the surface of the water … ought to have electric currents running directly across the path of her motion,' and the extensive contemporary researches into magnetism, Turner subtly links the imagery of iron filings on paper in an invisible magnetic field (fig. 128) with energised sea water surrounding and acting upon a magnetic iron ship. But in considering it, we should perhaps accept that this was not originally Turner's idea, but Thomas Campbell's in conversation within the same circle of scientists, who passed it on to Turner in his own words, either in talk or through his published *Poetical Works* (1836), a proof copy of which Turner owned:

There is a magnet-like attraction in
These waters to the imaginative power
That links the viewless with the visible,
And pictures things unseen.

NOTES

For abbreviated references see p. 136.

Place of publication is London, unless otherwise stated.

INTRODUCTION (pp. 9–11)

1 Gage 1980, nos. 318, 323.

2 It may have been shown in one of the sequences of uncatalogued displays in Turner's Gallery in Queen Anne Street in the 1810s and 1820s.

3 Gage 1980, p. 280.

4 Hamilton 1997, p. 273.

5 TB CXXXV.

CHAPTER I (pp. 12–20)

1 Royal Society Journal Books: Soane and Holland, 11 June 1795; Laporte, Varley and Jackson, 15 Feb. 1798.

2 Ibid., 26 Feb. 1801. The Revd Robert Nixon was an amateur painter and friend of Turner. See Hamilton 1997, pp. 20, 39–41; Bailey 1997, pp. 35–6.

3 Royal Society Journal Books, 20 Feb. 1794.

4 Quoted in D.G.C. Allen, *William Shipley: Founder of the Royal Society of Arts*, 1979, pp. 6–7.

5 The Society, now the Royal Society of Arts, still occupies these buildings, 8 John Adam Street.

6 John Gage, 'Turner and the Society of Arts', *Journal of the Royal Society of Arts*, vol. III, 1962–3, pp. 842–6. Hamilton 1997, pp. 34–5, medal reproduced between pp. 174 and 175.

7 Royal Institution Managers' Minutes, vol. I, March 1799–March 1800, p. 1, 9 March 1799. Published by the Royal Institution and Scolar Press [1972].

8 Egremont became a manager in 1799, and Fuller an annual subscriber in 1800.

9 John Paris, *The Life of Sir Humphry Davy*, 1830, p. 194.

10 See I. Morus, S. Schaffer and J. Secord, 'Scientific London', in *London: World City 1800–1840* (ed. C. Fox), Yale and London 1992, pp. 129–42.

11 Davy 1840, III, p. 2.

12 Babbage to Herschel, ?20/23 June 1831, Royal Society Library, HS2.262.

13 Edgeworth to Miss Buxton, 17 Jan. 1822. Som. Pap. c 370, MSE-1.

14 Somerville 1834, 3rd ed. 1836, p. 1.

15 H. Bence Jones, *Life and Letters of Faraday*, 1, 1870, pp. 377–8; also E.W. Cooke's Diary, 18 Jan. 1831 (Cooke Family Collection).

16 Thornbury 1877, pp. 349 et seq.

17 Chantrey to Babbage, 13 April 1832, Babb. Corr., Add. MS 37186, f. 331. Also M[ary] L[loyd], *Sunny Memories*, part 1, 1879, p. 50. And see Gage 1980, p. 280.

18 Babbage to Herschel, Dec. 1832, Royal Society Library, HS2.275.

19 R. Owen, *The Life of Richard Owen*, 1894, 1, pp. 262–4.

20 T.S. Cooper, *My Life*, 1890, quoted Finberg 1961, p. 432.

21 Faraday Collection. See Gertrude Prescott, 'Faraday: Image of the Man and the Collector', in D. Gooding and F.A.J.L. James (eds.), *Faraday Rediscovered*, 1985, pp. 15–31.

22 Babbage to D. Turner, 12 March 1836, DTP.

23 Larry J. Scharf, 'The Poetry of Light: Herschel, Art and Photography', in *John Herschel 1792–1871*, 1992, pp. 77–99.

24 Mary Somerville, Autobiography MS, first draft, Som. Pap. c 355, MSAU-2, f. 203. In the published version of Somerville's *Recollections* (1873) this paragraph is heavily edited.

25 M[ary] L[loyd], op.cit., part 2, 1880, pp. 31 et seq. Reprinted in *Turner Studies*, vol. 4, no. 1, 1984, pp. 22–3.

26 Thornbury 1877, pp. 348–52. Also see L.L. Reynolds and A.T. Gill, 'The Mayall Story', *History of Photography*, vol. 9, no. 2, 1985, pp. 89–107.

27 These have not been traced. A putative portrait, which I doubt, is reproduced in R.J.B. Walker, 'The Portraits of J.M.W. Turner', *Turner Studies*, vol. 3, no. 1, 1983, p. 30.

28 Hamilton 1997, p. 305; Bailey 1997, pp. 387–8.

29 Reference to inventory in Wilton 1987, p. 248; see also Hamilton 1997, pp. 211–13.

30 DTP, inside back flap of volume for Oct.–Dec. 1841; size of impression 7 × 5 mm.

31 P.W. Clayden, *Rogers and his Contemporaries*, 1887, II, p. 127, and Thornbury 1877, p. 315.

CHAPTER 2 (pp. 21–36)

1 W 1–3.

2 Thornbury 1877, pp. 26–7.

3 Hamilton 1997, chap. 1; Bailey 1997, p. 11.

4 See Thornbury 1877, pp. 30–2.

5 Thomas Sandby papers, RIBA Library, SaT/1/1&2.

6 *Oxford* sketchbook (TB II), p. 26a.

7 As listed between W 10 and W 262.

8 For example, *Bristol and Malmesbury* sketchbook (TB VI), details of Bath Abbey, p. 13.

9 W 16 and W 52.

10 *Oxford* sketchbook (TB II), pp. 6a–7.

11 W 333.

12 W 335–59.

13 Hamilton 1997, pp. 100, 209 and passim.

14 TB CXIV and CXXIX.

15 See Patrick Youngblood, 'The Painter as Architect: Turner and Sandycombe Lodge', *Turner Studies*, vol. 2, no. 1, 1982, pp. 20–35.

16 TB CXIV, pp. 78–77a.

17 Ibid.

18 The five and four digit figures are the numbers of the banknotes that Turner paid over. TB CXXVII, p. 3.

19 Hamilton 1997, pp. 134–8; Bailey 1997, chap. 13.

20 Hamilton 1997, chap. 5 and passim; Bailey 1997, chap. 12.

21 Studies of the East Lodge Gates at Farnley are in the *Farnley* sketchbook (TB CLIII), pp. 13a, 14.

22 '[£]14 – Jones for Water Closet', TB CCXI, inside front cover.

23 A small oil painting by George Jones RA, *Interior of Turner's Gallery: The Artist Showing his Works* (late 1840s, Ashmolean Museum, Oxford) gives a good idea of the layout of the gallery. Repr. Hamilton 1997, facing p. 175.

24 Plan, Howard de Walden Estate Contracts, vol. 4, p. 422.

25 BL Add. MS 46151-K, f. 13.

26 For example TB CXIV, pp. 53a–54, 66.

27 Maurice Davies, *Turner as Professor: The Artist and Linear Perspective*, exh. cat., Tate Gallery 1992.

28 Bound with his notes into BL Add. MS 46151-v and x.

29 Westminster Record Office, St Marylebone Ratebooks.

30 Thornbury 1877, p. 281; Derek Lindstrum, *Sir Jeffry Wyatville*, Oxford 1972, pp. 42 et seq.; Gage 1980, no. 149.

31 Cockerell Diaries, 14 Nov. 1821, RIBA Library.

32 Ibid., 2 Feb. 1825.

33 Cecilia Powell, 'Turner's "Antiquities at Pola": "the art of construction practically arranged"', *Turner Studies*, vol. 4, no. 1, 1984, pp. 39–43.

34 Cecilia Powell, 'Topography, Imagination and Travel: Turner's Relationship with James Hakewill', *Art History*, vol. 5, no. 4, 1982, pp. 408–25.

35 *Route to Rome* sketchbook, TB CLXXI. See also Hamilton 1997, p. 196.

36 Gage 1980, nos. 94, 127.

37 Turner to Holworthy, endorsed 7 Jan. 1826, Gage 1980, no. 112.

38 J. Davies to Robert Finch, 18 Nov. 1822. Finch Papers, d.5, Corresp. D-E, Bodleian Library, Oxford.

39 T.L. Donaldson to Robert Finch, 12 April 1824, loc. cit.

40 TB CCLIX 135.

41 J.M.W. Turner to Hawksworth Fawkes, 27 Dec. 1850, Gage 1980, no. 318.

42 J.M.W. Turner to Hawksworth Fawkes, 31 Jan. 1851, Gage 1980, no. 323.

CHAPTER 3 (pp. 37–57)

1 The *London Directory* of 1790 lists twenty-one coach makers, 4 ironmongers and braziers, one coach-lantern maker and one engine maker in Long Acre alone. Listed also in the street are glass grinders, trunk makers, livery lacemen, gold beaters, colourmen, saddlers, haberdashers, cabinet makers, japanners and lace and fringe makers.

2 Curtis Price, 'Turner at the Pantheon Opera House', *Turner Studies*, vol. 7, no. 2, 1987, pp. 2–8.

3 *Bristol and Malmesbury* sketchbook (TB VI), and W 15–21 and W 24–6.

4 *Dover Subjects* sketchbook (TB XVI B).

5 TB XVI G.

6 Transactions of the Society of Arts, 1792 and 1793; list of premiums, *Gents Magazine*, vol. 62, 1792, pp. 338–44.

7 TB IX A. For the context, see Hamilton 1997, pp. 30–31.

8 *Marford Mill* sketchbook (TB XX), pp. 11, 18 and passim.

9 TB XXVII N.

10 W 130 and W 169, also *Liber Studiorum* drawing and plate, TB CXVI O and British Museum 1945-12-8-323 respectively. See Forrester 1996, pp. 59–60.

11 TB XXVII P.

12 This was the pace of a slow hammer. The noise in more advanced forges was fearful – the Forge de Buffon, near Montbard in Burgundy, had a water-powered multiple trip hammer that struck 600 times a minute.

13 See Michael Kitson, 'Turner and Rembrandt', *Turner Studies*, vol. 8, no. 1, 1988, pp. 2–19.

14 W.G. Rawlinson, *The Engraved Work of J.M.W. Turner RA*, II, 1913, p. 374.

15 G.G.L. Hayes, Introduction, in Chris Evans (ed.), *The Letterbook of Richard Crawshay 1788–1797*, Cardiff 1990. Also John P. Addis, *The Crawshay Dynasty*, Cardiff 1957.

16 Anthony Bacon the younger was illegitimate and during his youth took his mother's surname, Bushby. He assumed the name of Bacon in 1792. See L.B. Namier, 'Anthony Bacon, MP, an Eighteenth-Century Merchant', *Journal of Economic and Business History*, vol. 2, 1929–30, pp. 20–70.

17 Cyfarthfa Papers, National Library of Wales.

18 Bacon is listed in poll books as Anthony Bacon, Esq., of Wantage in 1818, and of Benham House near Newbury in 1820. In his will, proved 1827, he is described as Anthony Bacon of Aberaman, Glamorgan: Berkshire Record Office, D/EX 1282/1 and D/EZ 43/7.

19 TB XXXVIII. Comparison with a letter from Anthony Bacon dated 22 Jan. 1794 in the Maybery Archive (2477; National Library of Wales) shows the handwriting beyond doubt to be Bacon's.

20 But see studies for *Vale of Ashburnham* in which most of the foreground detail is in place: *Vale of Heathfield* sketchbook (TB CXXXVII), pp. 68a–69.

21 TB XXXVII, p. 8.

22 Ibid., pp. 30–7.

23 Ibid., p. 82–3.

24 Lindsay 1966, p. 255.

25 TB XLII, pp. 26–9.

26 *Scotch Lakes* sketchbook (TB LVI), pp. 181a–182.

27 *Scotch Figures* sketchbook (TB LIX), passim.

28 *Hurstmonceux and Pevensey* sketchbook (TB XCI), pp. 62, 63, 80.

29 *Hesperides I* sketchbook (TB XCIII), p. 22a; *River* sketchbook (TB XCVI), p. 51a.

30 *River* sketchbook (TB XCVI), pp. 40, 54, 65a, 66a.

31 Ibid., pp. 13, 13a.

32 *Windsor and Eton* sketchbook (TB CVII), p. 67a.

33 The debate on the Pig Iron Duty Bill was reported at length in the *Times*, 10 May 1806.

34 *Harvest Home* sketchbook (TB LXXXVI). The oil painting is in the Tate Gallery (N00562), as is *Cassiobury Park: Reaping* (N04663).

35 *Farington's Diary*, vol. X, 7 July 1809; see also 5 July 1809.

36 Robert Hunt, in the *Examiner*, 15 May 1808; quoted B&J 81.

37 Lady Eastlake, *Diaries*, 1895, I, pp. 188–9.

38 The *Colour Bill* sketchbook (TB LXIII), contains some invoices for painting materials supplied by J. Newman, dated 1801.

39 TB CXXXV.

40 *Rosehill Park, Sussex*; B&J 211.

41 R.V. Saville, 'Gentry Wealth on the Weald in the Eighteenth Century: The Fullers of Brightling Park', *Sussex Archaeological Collections*, vol. 121, 1983, pp. 129–47.

42 W.R. Beswick, P.J. Broomhall and J.D. Bickersteth, 'Ashburnham Blast Furnace: A Definitive Date for its Closure', *Sussex Archaeological Collections*, vol. 122, 1984, pp. 226–7.

43 W 423–9, 431–5, 438.

44 Hamilton 1997, pp. 142–5, 218–19; Shanes 1990.

45 Tate Gallery (bequeathed by Henry Vaughan to the National Gallery 1900 and incorporated into the Turner Bequest, TB CXVIII R). See also Forrester 1996, no. 76.

46 W 426.

47 I am grateful to Roger Penhallurick of the Royal Cornwall Museum for information and for the Thorn photograph.

CHAPTER 4 (pp. 58–73)

1 M[ary] L[loyd], *Sunny Memories*, part 2, 1880, pp. 31–8, reprinted in *Turner Studies*, vol. 4, no. 1, 1984, pp. 22–3.

2 *Hastings* sketchbook (TB CXXXIX), p. 28a, *c*.1815–16.

3 I am grateful to Dr Margaret Penston of the Royal Greenwich Observatory, Cambridge, for clear information about eclipses in the early nineteenth century.

4 Thornbury 1877, p. 116; Hamilton 1997, p. 149.

5 TB CXIII, p. 48a.

6 Anaxagoras's contemporary Agatharcus was a painter whose decorative perspectives became legendary. The two are linked in a passage in Vitruvius, *De Architectura*, bk 7, praef. 11. This may have been Turner's reading when he made this note. Anaxagoras is also referred to by Plato in *Apology* and *Phaedo*. I am grateful to Professor Desmond Costa of the University of Birmingham for these references.

7 Wilton 1987, p. 248.

8 Samuel Rogers, 'The Pleasures of Memory', Pt II, from *Poems*, 1834, p. 33.

9 Finberg 1961, p. 252–3.

10 Pye MS, 86FF.73, f. 89 (National Art Library, Victoria and Albert Museum).

11 I am grateful to Dr John Thornes of the University of Birmingham for pointing out this weather effect to me.

12 Finley 1980, p. 178.

13 Thornbury 1877, pp. 138–9.

14 *Edinburgh Encyclopedia*, 1830, vol. 16, pp. 263 et seq. Quoted in Hamilton 1997, p. 216.

15 W.S. Dowden (ed.), *The Journal of Thomas Moore*, Delaware 1983, I, p. 257 (15 Nov. 1819).

16 First published in Davy 1840, I, p. 185.

17 Humphry Davy notebooks, 14g, pp. 5–7, Royal Institution, London.

18 Davy 1830, Dialogue the First, pp. 46–7, 55–6.

19 The date is not given in the published version of the vision. In Davy's notebooks, however, it is specified as having taken place in the autumn and early winter of 1818–19. Humphry Davy notebooks, 14h, pp. 170 et seq., Royal Institution, London. Davy recorded a subsequent vision in the Colosseum on 4 Nov. 1819; ibid., 21d, pp. 174–7.

20 Humphry Davy notebooks, 14g, p. 44.

21 Hamilton 1997, pp. 198–200.

22 Humphry Davy notebooks, 14g, p. 45.

23 Royal Academy of Arts, Lawrence letters LAW/3/39 Lawrence to Farington, 14 May 1819 [1818 written in error].

24 No. 19, 1807, reprinted Davy 1840, VIII, pp. 306–8.

25 Ibid.

26 Royal Society, 14 Jan. 1815.

27 For a full account of this aspect of Canova's career see Katharine Eustace, *Canova Ideal Heads*, Ashmolean Museum, Oxford 1997.

28 Phillips to D. Turner, 23 Feb. 1815, DTP.

29 TB CLXXIX, pp. 13a–21.

30 Somerville 1873, p. 269.

31 'Brief History of the Somerville Family', MS by William Somerville, 13 Sept. 1850, Som. Pap. MSFP-45.

32 Som. Pap. c. 369, 'Celebrities I' (Autographs).

33 'List from memory of pictures belonging to Mrs Somerville & her daughters in Cranley Place, no. 5;' dated Jan. 1872, Som. Pap. MSFP-56.

34 Forrester 1996, nos. 5, 43.

35 Turner's library is listed in Wilton 1987, pp. 246–7.

36 Gage 1987, pp. 222–4.

37 Samuel Rogers, 'The Campagna of Florence', lines 96–103; from *Italy*, 1830.

38 Mary Somerville, *Mechanism of the Heavens*, 1831, p. vi.

39 Canto I, verse xxvii. This connection was first traced by Butlin and Joll (see B&J 354) and enlarged upon by John Gage (1987, pp. 224–7).

40 Somerville 1834, p. 251.

41 Davy 1830, p. 22.

42 Royal Academy of Arts, Lawrence letters, LAW/2/108, 160: Lawrence to Davy, 1815 and ?1816.

43 Gage 1987, p. 224.

44 Gage 1980, no. 205.

45 *Reminiscences of Solomon Hart RA*, 1882, pp. 94–5.

CHAPTER 5 (pp. 74–91)

1 For a full account of Turner's painting see Judy Egerton, *Turner: The Fighting Temeraire*, 1995. See also Hamilton 1997, pp. 282–4; Bailey 1997, pp. 343–6.

2 Nigel Kennedy, *Records of Early British Steamships*, 1933, p. 7.

3 Turner to David Roberts [?1840s]; 'Henry Bicknell Collection of English Artists: Engraved Portraits, Letters etc', Paul Mellon Collection, Yale Center for British Art, p. 17. Published in Andrew Wilton, 'Turner as a Letter Writer', *Apollo*, vol. 113, 1981, p. 59, and John Gage, *Turner Studies*, vol. 6, no. 1, 1986, p. 8.

4 Davy 1830, p. 34.

5 T.L. Donaldson to Robert Finch, 14 Aug. 1824, Finch Papers, d. 5, Corresp D–E, ff. 150–1, Bodleian Library, Oxford; Thomas Phillips to D. Turner, 17 Nov. 1829, DTP.

6 See Shanes 1990, pp. 71, 122–3, 132, 147; and pp. 203, 273.

7 See William S. Rodner, 'Turner and Steamboats on the Seine', *Turner Studies*, vol. 7, no. 2, 1987, pp. 36–41.

8 Turner to James Lenox, 16 Aug. 1845; Gage 1980, no. 288.

9 Walter Scott, 'Lord of the Isles', canto iv, verse x.

10 Anon, 'The Poetical Works of Thomas Campbell', *Quarterly Review*, vol. 57, Dec. 1836, pp. 349–61. William Rodner (op. cit.) has identified the reviewer as William Henry Smith.

11 Thomas Campbell, 'Lines on the View from St Leonards,' 1831, lines 75–6. These are quoted more fully in Chapter 7.

12 David Blayney Brown, 'Rule, Britannia? Patriotism, Progress and the Picturesque in Turner's Britain', in *Turner*, exh. cat., National Gallery of Australia 1996, pp. 48–72.

13 'Captain Manby's Invention', *Times*, 4 May 1812.

14 George William Manby, *A Lecture on the Preservation of Persons in the Hour of Shipwreck*, 1814.

15 'Captain Manby's Apparatus for Wrecks', *Edinburgh Review*, vol. 76, May 1823, pp. 333–49.

16 *Faraday Letters*, 1, nos. 442, 443.

17 See also Kenneth Walthew, *From Rock and Tempest: The Life of Captain George William Manby*, 1971.

18 *Gentleman's Magazine*, May 1831, pp. 451–2.

19 As listed in the *Edinburgh Encyclopaedia*, 1830, entry for 'Lighthouses'.

20 Walter Scott wrote a letter introducing Turner to Stevenson on 30 April 1819 and Turner wrote to Stevenson on 8 July 1819 saying he had despatched the watercolour on 'Saturday last'. See Finberg 1961, p. 257, and Gage 1980, nos. 84, 85, and p. 287.

21 *Wilson* sketchbook (TB XXXVII), p. 8; TB XXVII T (*A Lighthouse from the Land*, 1790s.)

22 TB CXCVIII, pp. 8, 10a, 15, 69.

23 Sept. 1840, p. 384.

24 Ruskin to Wethered, 12 Dec. 1843, copy in Victoria and Albert Museum, 86.FF.73, f. 10.

CHAPTER 6 (pp. 92–114)

1 Lindsay 1966, p. 152; Eric Shanes, *Turner's Human Landscape*, 1990, pp. 231–8.

2 One survives at Morwhellam Quay near Tavistock. Turner made studies of industrial workings during his 1811 and 1812 Devon tours, e.g. *North Devon* sketchbook (TB CXXVA), pp. 51–6.

3 A full analysis of Turner's *Leeds* is given in Stephen Daniels, 'The Implications of Industry: Turner and Leeds', *Turner Studies*, vol. 6, no. 1, 1986, pp. 10–17.

4 See Hamilton 1997, pp. 94–7, 110–13 and passim; Bailey 1997, chap. 12.

5 TB CC.

6 'Hurries' is the local name given to coal chutes and allied coal-wharf equipment.

7 TB LXXXIX, p. 14.

8 See B&J 356, 360.

9 See Ian Warrell, 'Lord of the Soil: A Re-appraisal of Turner's Petworth Patron'; Appendix 1 in Martin Butlin, Mollie Luther and Ian Warrell, *Turner at Petworth: Painter and Patron*, 1989, pp. 105–23. For a very jaundiced view of the future of the canal see the *Times*, 17 Nov. 1827.

10 See 'Wall Plan of the East Side of the Carved Room', ibid., p. 116.

11 See Andrew Wilton and Rosalind Turner, *Painting and Poetry: Turner's Verse Book and his Work of 1804–1812*, exh. cat., Tate Gallery 1990.

12 I have suggested (in Hamilton 1997, p. 231), that Turner might have held shares in the Chichester Canal. On later reflection, however, I now believe that the shareholder listed as 'William Turner' was someone else altogether.

13 See the commentary to this picture in Shanes 1990, p. 237. A full account of *Dudley* is given in William S. Rodner, 'Turner's *Dudley*: Continuity, Change and Adaptability in the Industrial Black Country', *Turner Studies*, vol. 8, no. 1, 1988, pp. 32–40.

14 L.T.C. Rolt, *Isambard Kingdom Brunel*, 1957, new ed. 1989, p. 57.

15 Ibid.

16 Comparison with the anonymous painting *The Banquet in the Thames Tunnel* (private collection) is instructive: this is illustrated in Francis Klingender, *Art and the Industrial Revolution*, 1968.

17 But see reference to 'Mr Turner' in connection with the Horsley/Calcott/Brunel circle in Rosamund Brunel Gotch (ed.), *Mendelssohn and his Friends in Kensington*, Oxford 1934, p. 261.

18 Hilaire Faberman and Philip McEvansoneya, 'Isambard Kingdom Brunel's "Shakespeare Room"', *Burlington Magazine*, vol. 137, Feb. 1995, pp. 108–18.

19 Extracts from Brunel's Journals: 'Returned to town by 8 o'clock coach. Walk with B-n to Westmacott's, Chantrey's etc etc dined with them' (21 Sept. 1829); 'Dined at Athenaeum with Dr Somerville & party. Walked with Alexander – returned home' (15 July 1830); 'Went to [Decimus] Burton's soirée. Walked home with Babbage' (9 June 1831): Brun. Pap. DM 1306 / 1.3.iii and 11.3.i.

20 Hamilton 1997, pp. 232–3.

21 Rolt op. cit., p. 142. Brunel Journals, 26 Dec. 1835. Brun. Pap. DM 1306 / 11.3.ii.

22 *Middlesex Journal and West Surrey Gazette*, 4 Feb. 1840. See also Sir Alfred Pugsley (ed.), *The Works of Isambard Kingdom Brunel*, 1976, pp. 91–106.

23 For further views on the painting see John Gage, *Turner: Rain, Steam, and Speed*, Art in Context series, 1972; John McCoubrey, 'Time's Railway: Turner and the Great Western', *Turner Studies*, vol. 6, no. 1, 1986, pp. 33–9; Hamilton 1997, p. 300; Bailey 1997, pp. 363–5.

24 B&J 523.

25 Hamilton 1997, pp. 253–4.

26 *Gentleman's Magazine*, vol. 101, Aug 1831, pp. 121–9.

27 See B&J 346.

28 There are a number of studies of the old London Bridge in the *Old London Bridge* sketchbook (TB CCV) and the *London Bridge and Portsmouth* sketchbook (TB CCVI); studies of the new bridge under construction in the *Isle of Wight* sketchbook (TB CCXXVII), pp. 12 and 32a; and of the completed bridge in the *Edinburgh* sketchbook (TB CCLXVIII), p. 63, and *Mouth of the Thames* sketchbook (TB CCLXXVIII), pp. 8a–9.

29 *Gentleman's Magazine*, vol. 101, Aug. 1831, p. 129.

30 Hamilton 1997, pp. 296–7; Peter Bicknell and Helen Guitermann, 'The Turner Collector: Elhanan Bicknell', *Turner Studies*, vol. 7, no. 1, 1987, pp. 34–44.

31 Robert K. Wallace, 'The Antarctic Sources for Turner's 1846 Whaling Oils', *Turner Studies*, vol. 8, no. 1, 1988, pp. 20–31.

32 Ibid.; see also Bicknell and Guitermann, loc cit. n. 29.

33 Hamilton 1997, pp. 296–7.

34 James Clark Ross, *A Voyage of Discovery and Research in the Southern and Antarctic Regions during the Years 1839–1843*, 2 vols., 1847.

35 Gage 1980, no. 255.

36 B&J 401.

37 B&J 404, 405.

38 R. Owen, *The Life of Richard Owen*, 1894, I, pp. 262–4. See also Hamilton 1997, pp. 272–3; Bailey 1997, pp. 272, 379. For Broderip, see 'W. J. Broderip: In Memoriam', *Fraser's Magazine*, vol. 59, 1859, pp. 485–8.

39 Owen, op. cit., I, pp. 262–4.

40 Richard Owen to Lubbock, 20 April 1839, Royal Society, LUB. O. 66.

41 Owen, op. cit., I, p. 264.

42 For Turner's illnesses at this period see Hamilton 1997, chap. 14, and Appendices 2 and 3.

43 Andrew Wilton has put forward the view that the equestrian statue of Wellington is not Wyatt's, but Chantrey's, cast after the sculptor's death by Henry Weekes and erected in 1840 outside the Royal Exchange: Wilton 1979, pp. 227–8, n. 107.

44 Wyatt's workshop ledgers are in the RIBA Library.

45 Two articles examine different aspects of the background to Turner's painting: Nicholas Alfrey, 'Turner and the Cult of Heroes', *Turner Studies*, vol. 8, no. 2, 1988, pp. 33–44; and John McCoubrey, '*The Hero of a Hundred Fights*: Turner, Schiller and Wellington', *Turner Studies*, vol. 10, no. 2, 1990, pp. 7–11.

46 *Illustrated London News*, 6 June, 15 Aug., 3 Oct., 21 Nov. 1846.

47 *Illustrated London News*, 3 Oct. 1846, p. 224.

48 *Literary Gazette*, 13 Sept. 1845, p. 612.

49 John McCoubrey, loc. cit. n. 44.

50 Book of Revelation, xix:17–18.

51 Davy 1830, p. 55.

CHAPTER 7 (pp. 115–128)

1 This is based on the watercolour in the *St Gothard and Mont Blanc* sketchbook (TB LXXV), p. 33.

2 18 April 1815, DTP.

3 Turner to Buckland, 30 Nov. 1843, Gage 1980, no. 262.

4 T. Dickinson, *A Narrative of the Operations of the Recovery of the Public Stores and Treasure Sunk in HMS Thetis at Cape Rio, on the Coast of Brazil on the 5th December 1830*, 1836, pp. 61 et seq. Quoted Gage 1980, no. 262, no. 5.

5 See letters between Chantrey and Buckland, 7 Jan. 1826 and 8 March 1844, Osborn Collection, Beinecke Library, Yale; see also William Buckland, *Geology and Mineralogy Considered with Reference to Natural Theology: The Bridgewater Treatise*, IV, 1837, p. 305n.

6 Clara Wells writing to Thornbury (Thornbury 1877, p. 236). See also Gage 1987, pp. 219–20.

7 Hamilton 1997, pp. 147–8 and passim. Portrait reproduced between pp. 174 and 175.

8 Thornbury 1877, p. 236.

9 Luke Herrmann, 'Ruskin and Turner: A Riddle Resolved', *Burlington Magazine*, vol. 112, Oct. 1970, pp. 696, 699.

10 George Jones, *Sir Francis Chantrey, RA: Recollections of his Life, Practice and Opinions*, 1849, pp. 98 et seq.

11 See Forrester 1996, passim, and Appendix B.

12 TB CXXIII, p. 58v. See Hamilton 1997, pp. 143–5.

13 Ibid., p. 149v.

14 Ibid., p. 201v.

15 Ibid., p. 140v.

16 There is, however, no record in the Geological Society minutes that Turner was ever a guest at a meeting.

17 *Transactions of the Geological Society*, vol. I, pp. 93–184, 248–68.

18 Gage 1987, pp. 219–20. Though listed by Bernard Falk in his *Turner the Painter: His Hidden Life*, 1938, Appendix 1, pp. 255–9 (as '2 vols. 4to. 1811') these volumes were missing when Wilton came to list Turner's library in Wilton 1987, pp. 246–7. Falk's citation is not completely accurate, as the two volumes were published in 1811 and 1814 respectively.

19 BL Add. MS 46151-s, f. 10.

20 See Finley 1980.

21 Sir Walter Scott, 'Lord of the Isles', *Poetical Works*, 1834, Canto III, verse xiv.

22 See also Sir David Brewster, *Letters on Natural Magic Addressed to Sir Walter Scott, Bart*, 1832.

23 John McCulloch, *The Highlands and Western Isles of Scotland*, 1824, III, pp. 474–84.

24 See B&J 132. The subject was engraved by Charles Turner in Nov. 1815. The one surviving impression is in the British Museum.

25 An effect which came to be known by photographers in the late 1850s as 'halation'. Seen also in Turner's *Chichester Canal* (c.1828, Tate Gallery) and *Mortlake Terrace* (exh. 1827, National Gallery of Art, Washington.)

26 Somerville to Revd Dr Somerville; Naples, 24 March 1818; Som. Pap. c 360, MSFP-44.

27 Possibly by James Hakewill.

28 This is untraced. See Gage 1980, p.92n; *Art Journal*, vol. 7, 1868, p.129. While on the subject of Vesuvius, I must scotch a myth. Gage 1969, p.131 (echoed by Bailey 1997, p.246) says that Turner 'offered up his sketchbook to receive the mountain's signature, which came as a shower of hot ash'. Delightful though this idea is, it did not happen. The marks on these pages (TB CLXXXIV, pp.78a–79 and p.82) are of a light inky paint splashed onto them later in the studio.

29 For the contemporary issues see Nicholaas A. Rupke, *The Great Chain of History*, Oxford 1983, and Anthony Hallam, *Great Geological Controversies*, Oxford 1989, 2nd ed.

30 Erasmus Darwin, *Temple of Nature*, 1803, IV, lines 35–6. See also Gage 1969, p.129.

31 Faraday to Babbage, 12 Feb. 1840, *Faraday Letters*, II, no.1240.

32 Chantrey to Faraday, 19 Jan. 1836, ibid., no.874.

33 Cockerell Diaries, 13 Dec. 1825, RIBA Library.

34 C.R. Leslie, *Autobiographical Recollections*, ed. T. Taylor, 1860, I, p.77.

35 Humphry Davy notebooks, 14g, p.44, Royal Institution.

36 W.B. Pope (ed.), *The Diary of Benjamin Robert Haydon*, 1960, III, 3–16 June 1829.

37 Somerville 1873, p.137. Somerville Island lies in Lancaster Sound between Somerset and Cornwallis Islands in the N.W. Territories of Canada: 74° 44' N – 96° 11' W.

38 Mary Somerville: 'On the magnetizing power of the more refrangible solar rays.' Communicated to the Royal Society by Dr William Somerville, 2 Feb. 1826, *Phil. Trans*, vol. 116, 1826. Mary Somerville was not permitted to deliver her paper to the Royal Society herself because she was a woman. For her admission that she later destroyed copies of the paper see Som. Pap. c.355, MSAU-2, p.104. See also E.C. Patterson, *Mary Somerville and the Cultivation of Science 1815–1840*, The Hague 1983, pp.45–8. Also Thornbury 1877, pp.349 et seq.; Gage 1987, pp.222 et seq.

39 'Experimental Researches in Electricity: 1st Paper.'

40 Faraday refers to them in a note in *Experimental Researches in Electricity* (Series 1, Nov. 1831), vol. 1, 1849, p.32

41 See Gage 1980, nos. 230, 242, 277, 303, 306, 335; also nos. 241, 248, 269, 292.

42 Royal Institution Managers' Minutes, vol. 12, p.32 (21 Nov. 1863) and p.84 (5 Dec. 1864).

43 She is the woman in white with a lace bonnet and wearing glasses, half way up on the left-hand edge, between two women in black. Key to lithograph in Royal Institution. See also letter, Tyndall to Moore [1850s], *Tyndall Correspondence*, III, p.863, Royal Institution.

44 J. Bence Jones, *The Life and Letters of Faraday*, 1870, I, p.378.

45 C. Hullmandel, *The Art of Drawing on Stone*, 1824, p. vii.

46 Thomas Campbell, 'Lines on the View from St Leonard's, 1831', lines 73–6, 82–9.

47 Somerville 1834, pp.352–3.

48 George Biddle Airy, 'Account of Experiments on Iron-built ships, instituted for the purpose of discovering a correction for the deviation of the compass produced by the iron of the ships'; read to the Royal Society, 25 April 1839, concluded 9 May 1839. Abstract published *Phil. Trans.* vol. 4, 1837–43, pp.141–5.

49 S. Hunter Christie: 'Report on the State of our Knowledge Respecting the Magnetism of the Earth'; Report of the British Association for the Advancement of Science (Cambridge meeting, 1833), 1834.

50 See Hamilton 1997, p.290; Bailey 1997, pp.277–8.

51 Charles Ninnis has suggested that the '*Ariel*' is *The Fairy*, which was lost out of Harwich in a storm in November 1840; *Turner Society News*, no. 20, Jan. 1981, pp.6, 8. See also letter, G.W. Manby to D. Turner, 1 March 1841 (DTP), in which Manby writes of the loss of a friend 'in Her Majesty's Survey Ship Fairy.'

ABBREVIATIONS

Place of publication is London, unless otherwise stated.

Babb. Corr.	Babbage Correspondence, British Library
Bailey 1997	Anthony Bailey, *Standing in the Sun: A Life of J.M.W. Turner*, 1997
B&J	Martin Butlin and Evelyn Joll, *The Paintings of J.M.W. Turner*, revised ed., 2 vols. 1984
BL	British Library
Brun. Pap.	Brunel Papers, University of Bristol Library
Davy 1830	Humphry Davy, *Consolations in Travel, or the Last Days of a Philosopher*, 1830
Davy 1840	*The Collected Works of Sir Humphry Davy, Bart*, ed. John Davy, 1840
DTP	Dawson Turner Papers, Trinity College Library, Cambridge
Faraday Letters	*Letters of Michael Faraday*, ed. F.A.J.L. James, 2 vols. 1991 and 1993
Farington's Diary	*The Diary of Joseph Farington RA 1793–1821*, ed. Kenneth Garlick, Angus McIntyre and Kathryn Cave, 16 vols., New Haven and London 1978–84
Finberg 1961	A.J. Finberg, *The Life of J.M.W. Turner*, revised ed., Oxford 1961
Finley 1980	Gerald Finley, *Landscapes of Memory: Turner as Illustrator to Scott*, Los Angeles 1980
Forrester 1996	Gillian Forrester, *Turner's 'Drawing Book': The Liber Studiorum*, exh. cat., Tate Gallery 1996
Gage 1969	John Gage, *Colour in Turner: Poetry and Truth*, 1969
Gage 1980	John Gage, *Collected Correspondence of J.M.W. Turner*, Oxford 1980
Gage 1987	John Gage, *J.M.W. Turner: 'A Wonderful Range of Mind'*, New Haven and London 1987
Hamilton 1997	James Hamilton, *Turner: A Life*, 1997
Lindsay 1966	Jack Lindsay, *J.M.W. Turner: A Critical Biography*, 1966
Phil. Trans	*Philosophical Transactions of the Royal Society*
R	W.G. Rawlinson, *The Engraved Work of J.M.W. Turner, RA*, 2 vols., 1908 and 1913
RIBA	Royal Institute of British Architects, London
Shanes 1990	Eric Shanes, *Turner's England 1810–38*, 1990
Somerville 1834	Mary Somerville, *On the Connexion of the Physical Sciences*, 1834
Somerville 1873	Martha Somerville (ed.), *Personal Recollections from Early Life to Old Age of Mary Somerville, with Selections of her Correspondence*, 1873
Som. Pap.	Somerville Papers, Bodleian Library, Oxford
TB	A.J. Finberg, *A Complete Inventory of the Drawings of the Turner Bequest*, 2 vols., 1909
Thornbury 1877	Walter Thornbury, *Life and Correspondence of J.M.W. Turner*, 1877 ed.
Wilton 1979	Andrew Wilton, *The Life and Work of J.M.W. Turner* [with a catalogue of watercolours outside the Turner Bequest, cited as 'W'], 1979
Wilton 1987	Andrew Wilton, *Turner in his Time*, 1987

CATALOGUE

Works are by J.M.W. Turner,
and are on paper, unless otherwise
stated.

Measurements are given in
centimetres, followed by inches in
brackets, height before width.

When a figure illustration shows
only one page from a sketchbook
spread, this is identified by an
asterisk (*).

A dagger (†) indicates that the work
is displayed on the main floor

1 People and Ideas

Sir Francis Chantrey RA, FRS
(1781–1841)

1 *Bust of Mary Somerville* 1840 (fig. 7)
Marble 61 × 53.3 × 27.9
(24 × 21 × 11), socle 12.7 (5)
The President and Council of the
Royal Society

Matthew Noble (1818–1876)

2 *Bust of Michael Faraday* 1854 (fig. 8)
Marble 55.9 × 53.3 × 30.5
(22 × 21 × 12), socle 15.2 (6)
The President and Council of the
Royal Society

Michael Wagmüller (1839–1881)

3 *Bust of Sir Richard Owen* 1871
(fig. 10)
Marble 74 × 48 × 28
(29⅛ × 18⅞ × 11)
The Trustees of the Natural History
Museum

Sir Francis Chantrey RA, FRS
(1781–1841)

4 *Bust of John Fuller MP* 1820 (fig. 3)
Marble 78 × 52 × 35
(30¾ × 20½ × 13¾)
Royal Institution of Great Britain

Harriet Moore (1796/7–1884)

5 *The Magnetic Laboratory of Michael
Faraday c.*1850 (fig. 9)
Watercolour 17 × 24.6 (6¾ × 9⅝)
Royal Institution of Great Britain

Sir Humphry Davy, Bt, PRS
(1778–1829)

6 *Pages from Notebook 14h of Sir
Humphry Davy Containing his Account
of his Visions* 1819–20
Notebook 14 × 21.6 (35.6 × 8½)
Royal Institution of Great Britain

Mary Somerville (1780–1872)

7 *Mole de Gaeta* 1839 (fig. 11)
Oil 19.5 × 28 (7 5/8 × 11)
Edmund Fairfax-Lucy

Mary Somerville

8 *Italian Landscape: View towards a Bay*
1830s (fig. 12)
Oil 13.2 × 27.4 (5¼ × 10¾)
Edmund Fairfax-Lucy

2 Foundations

9 *Wanstead New Church* 1789–90
(fig. 14)
Pencil 21.6 × 31.8 (8½ × 12½)
TB IV A; D00051

10 *Part of the South Front of Fonthill
Abbey, the Tower in Process of Erection*
?1799 (fig. 18)
from dismembered *Fonthill*
sketchbook
Pencil 46.5 × 32.5 (18¼ × 12¾)
TB XLVII 1; D02178

11 *Fonthill Abbey – the Tower in the
Process of Erection* ?1799
Pencil 33.4 × 46.6 (13⅛ × 18⅜)
TB XLVII 5; D02182

12 *St Mary's and the Radcliffe Camera from
Oriel Lane, Oxford* mid 1790s (fig. 16)
Watercolour 52.7 × 38.5
(20¾ × 15⅛)
TB XXVII X; D00685

13 *Sketches and Calculations for
Sandycombe Lodge c.*1810 (fig. 19)
from dismembered *Sandycombe and
Yorkshire* sketchbook
Pen and ink 12.3 × 20 (4⅞ × 7⅞)
TB CXXVII, p. 13a; D08982

14 *Admiralty Screen and Building: Front
View c.*1810 (fig. 22)
Perspective diagram: lecture
illustration
Watercolour 78.1 × 132.1
(30¾ × 52)
TB CXCV 173; D17144

15 *Interior of a Prison c.*1810 (fig. 23)
Perspective diagram: lecture
illustration
Pen and wash 36.8 × 50.8
(14½ × 20)
TB CXCV 121; D17091

16 *Entablature of a Tuscan Column, with
Perspective Lines c.*1810 (fig. 24)
Perspective diagram: lecture
illustration
Pen and ink 66.7 × 100.4
(26¼ × 39½)
TB CXCV 92; D17062

17 *Study for 'Rome from the Vatican'* 1819
(fig. 29)
Pen and ink 36.8 × 22.8 (14½ × 9)
TB CLXXXIX 41; D16368

SKETCHBOOKS

18 *Early Perspective Study: North Front of
Radley Hall c.*1789 (fig. 15)
Oxford sketchbook
Pencil 16.2 × 24.7 (6⅜ × 9¾)
TB II, p.9; D00021

19 *Studies inside a Picture Gallery* 1808
    (fig. 21)
    Tabley No. 3 sketchbook
    Pencil 10.8 × 18.4 (4¼ × 7¼)
    TB CV, pp. 67a–68*; D07085–6

20 *Notes on Architectural Perspective* 1809
    (fig. 25)
    Perspective sketchbook
    Pen and ink 8.8 × 11.4 (3½ × 4½)
    TB CVIII, pp. 43a–44*; D07428–9

21 *Study of Fortifications at Laurenstein*
    1817 (fig. 28)
    Waterloo and Rhine sketchbook
    Pen and ink 14.9 × 9.5 (5⅞ × 3¾)
    TB CLX, p. 91a, D12879

22 *Landscape with Workmen Building a*
    Bridge c.1817 (fig. 26)
    Durham, North Shore sketchbook
    Pencil 11.4 × 18.7 (4½ × 7⅜)
    TB CLVII, p. 20; D12343

23 *Waterloo Bridge from Beneath*
    c.1818–20 (fig. 27)
    Scotland and London sketchbook
    Pencil and wash 11.3 × 18.7
    (4½ × 7⅜)
    TB CLXX, p. 6; D13821

24 *Notes of Building Work at Queen Anne*
    Street c.1820
    Paris, Seine and Dieppe sketchbook
    Pencil 11.3 × 18.7 (4½ × 7⅜)
    TB CCXI, inside cover; D40976

3 The Old Technology

25 *Watermill and Stream* 1791–2
    (fig. 31)
    Oil, oval 24.7 × 30 (9¾ × 11¾)
    TB XXXIII a; D00898

26 *Marford Mill, Denbighshire* 1794
    (fig. 30)
    Pencil 21.2 × 13.7 (8⅜ × 5⅜)
    TB XXI L; D00339

27 *Limekiln at Coalbrookdale* c.1797
    (fig. 36)
    Oil on panel 29 × 40.3
    (11⅜ × 15⅞)
    Yale Center for British Art, Paul
    Mellon Collection

28 *Cyfarthfa Iron Works* 1798
    (fig. 38)
    Pencil 45.7 × 29.3 (18 × 11½)
    TB XLI 1; D01629

29 *Cyfarthfa Iron Works* 1798
    (fig. 39)
    Pencil 45.7 × 29.3 (18 × 11½)
    TB XLI 2; D01630

30 *An Iron Foundry* 1798 (fig. 34)
    Watercolour 24.8 × 34.3
    (9¾ × 13½)
    TB XXXIII B; D00873

31 *Blacksmith's Shop* c.1806–10
    (fig. 43)
    Study for a *Liber Studiorum*
    engraving
    Sepia 19.5 × 27.3 (7⅝ × 10¾)
    TB CXVI H; D08109

32 *A Country Blacksmith Disputing upon*
    the Price of Iron, and the Price Charged
    to the Butcher for Shoeing his Poney
    exh. 1807 (fig. 42)†
    Oil on wood 54.9 × 77.8
    (21⅝ × 30⅝)
    Tate Gallery. Bequeathed by the
    artist 1856
    N00478

33 *An Artists' Colourman's Workshop*
    c.1807 (fig. 44)†
    Oil on wood 62.2 × 91.4
    (24½ × 36)
    Tate Gallery. Bequeathed by the
    artist 1856
    N05503

34 *Dorchester Mead, Oxfordshire*
    exh. 1810 (fig. 47)
    Oil on canvas 101.6 × 130.2
    (40 × 51¼)
    Tate Gallery. Bequeathed by the
    artist 1856
    N00485

George Cooke,
after J.M.W. Turner

35 *Tintagel Castle* 1818 (fig. 53)
    Engraving 16.2 × 24.1 (6⅜ × 9½)
    Tate Gallery. Purchased 1988
    R106; T05443

36 *Vale of Ashburnham* 1816 (fig. 48)
    Watercolour 38 × 56.4 (15 × 22¼)
    The British Museum

W.B. Cooke, after J.M.W. Turner

37 *Vale of Heathfield* 1818 (fig. 49)
    Engraving 19.1 × 28.4 (7½ × 11⅛)
    Tate Gallery. Purchased 1986
    T04436; R133

38 *Crowhurst* c.1816 (fig. 50)
    Pen and ink, pencil and
    watercolour 20 × 27.2 (7⅞ × 10¾)
    TB CXVIII R; D08172

Philippe Jacques de Loutherbourg
(1740–1812)

39 *Welsh & Shropshire Industrial Subjects:*
    'Large Fire Engine in Coalbrook Dale';
    'Fire Engin Coalbrook Dale'; 'Largest
    Fire-Engin of Coalbrook Dale' (fig. 37)
    Pen and ink on card,
    each 12.4 × 8 (4⅞ × 3⅛)
    TB CCCLXXII 45, 47*, 49; D36405,
    D364407, D36409

SKETCHBOOKS

40 *Nunwell and Brading from Bembridge*
    Mill 1795 (fig. 33)
    Isle of Wight sketchbook
    Watercolour 26.5 × 20.4 (10⅜ × 8)
    TB XXIV 49; D00458

41 *Forge Scene* 1796–7 (fig. 40)
    Wilson sketchbook
    Watercolour 11.3 × 9.4 (4½ × 3¾)
    TB XXXVII 102–3; D01219–20

42 *Foundry with Tilt Hammer* c.1798
    (fig. 41)
    Swans sketchbook
    Pen and ink 17.5 × 12.4 (6⅞ × 4⅞)
    TB XLII 60–1; D01735–6

43 *Interior of a Workshop* 1807–8 (fig. 45)
    River and Margate sketchbook
    Pencil 11.4 × 19.1 (4½ × 7½)
    TB XCIX inserted at p. 77; D06494

44 *Composition Studies (with Tripod and*
    Pulley) for 'Dorchester Mead,
    Oxfordshire' 1805–9 (fig. 46)
    Hesperides II sketchbook
    Pen and ink 22.8 × 14.6 (9 × 5¾)
    TB XCIV 4a–5*; D05850–1

45 *Tintagel Castle, with Lifting Gear* 1811
    (fig. 51)
    North Devon sketchbook
    Pencil 14 × 21.6 (5½ × 8½)
    TB CXXVa 32; D41308

4 Observing the Sky

46 *A Rainbow with Cattle* ?1816 (fig. 65)
Watercolour 25.8 × 41.4
(10¹⁄₈ × 16¹⁄₄)
TB CXCVII G; D17197

47 *Arundel Castle with Rainbow* c.1824
(fig. 66)
Watercolour 16 × 22.9 (6¹⁄₄ × 9)
TB CCVIII F; D18139

48 *Eclipse, with a Gesticulating Figure*
1824 (fig. 58)
Watercolour painted on the back
of an envelope addressed to
Turner, franked 5 June 1824
11.9 × 22.8 (4⁵⁄₈ × 9)
TB CCLXXX 139; D27656

49 *Sunset – Nancy. Study for 'The Rivers of
Europe'* early 1830s (fig. 63)
Watercolour and gouache
14 × 18.8 (5¹⁄₂ × 7³⁄₈)
TB CCLIX 192; D24757

50 *Small High Cloud Catching the Sunset
(Sunset at Sea)* 1830s (fig. 64)
Watercolour 29.3 × 39.4
(11¹⁄₂ × 15¹⁄₂)
TB CCLXIII 68; D25190

51 *Study for 'Galileo's Villa'* c.1826–7
(fig. 67)
Preparatory drawing for illustration
to Rogers's *Italy* (1830)
Pencil and watercolour 14.8 × 19.6
(5⁷⁄₈ × 7³⁄₄)
TB CCLXXX 87; D27604

52 *Greenwich Hospital* c.1831–2 (fig. 60)
Illustration to Rogers's *Poems*, 1834
Watercolour 19.1 × 25.6
(7¹⁄₂ × 10¹⁄₈)
TB CCLXXX 176; D27693

53 *The Mustering of the Warrior Angels*
c.1833–4 (fig. 68)
Illustration to 'Paradise Lost' in
Milton's *Poetical Works* (1835)
Watercolour 13.5 × 11.5
(4⁵⁄₈ × 4¹⁄₈)
Preston Hall Museum, Stockton on
Tees Museums Service

54 *Imaginative Sunset* mid 1830s (fig. 69)
Watercolour 17.9 × 22.6 (7 × 8⁷⁄₈)
TB CCLXXX 70; D27587

SKETCHBOOKS

55 *Study of a Partial Eclipse of the Sun*
1804 (fig. 55)
Eclipse sketchbook
Charcoal and chalk 16.5 × 10.8
(6¹⁄₂ × 4¹⁄₈)
TB LXXXV, pp.1a–2; D05246–7

56 *Brightling Observatory* 1815–16 (fig. 59)
Hastings sketchbook
Pencil 12.7 × 20.4 (5 × 8)
TB CXXXIX, p.34; D10393

57 *Skyscape over Hawksworth Moor, with
Colour Notes* 1816–18 (fig. 61)
Scarborough II sketchbook
Pencil 18.1 × 11.8 (7¹⁄₈ × 4⁵⁄₈)
TB CLI 18a; D11973

58 *Study of the Sky* 1818 (fig. 62)
Skies sketchbook
Watercolour 12.4 × 24.8 (4⁷⁄₈ × 9³⁄₄)
TB CLVIII, p.10; D12458
[illustration: TB CLVIII, p.54;
D12502]

59 *Schematic Memorandum of a Sunrise*
1822 (fig. 56)
King's Visit to Scotland sketchbook
Pencil 11.4 × 19.1 (4¹⁄₂ × 7¹⁄₂)
TB CC, pp.2a–3; D17511–12

60 *Diagrammatic Progress of the Setting Sun*
?1832 (fig. 57)
Life Class sketchbook
Pencil 8.6 × 11.2 (3³⁄₈ × 4³⁄₈)
TB CCLXXIX (a), 63a–64*;
D27478–9

5 From Sail to Steam

61 *The Bell Rock Lighthouse* 1819 (fig. 90)
Watercolour and gouache with
scratching out 30.6 × 45.5
(12 × 17⁷⁄₈)
National Gallery of Scotland,
Edinburgh

62 *Dartmouth on the River Dart* 1822
(fig. 74)
Watercolour 15.7 × 22.7 (6¹⁄₈ × 9)
TB CCVIII C; D18136

63 *Dover* c.1825 (fig. 75)
Watercolour 16.1 × 24.5 (6³⁄₈ × 9⁵⁄₈)
TB CCVIII U; D18154

J.T. Willmore,
after J.M.W. Turner

64 *Dover* 1851 (fig. 76)
Engraving, image size 40.6 × 59.5
(16 × 23³⁄₈)
Tate Gallery. Purchased 1990
R666; T05791

65 *Harbour with Figures and Shipping,
Margate* c.1826 (fig. 84)
Watercolour 14 × 18.8 (5¹⁄₂ × 7³⁄₈)
TB CCXXIV 29; D20319

66 *Lighthouse against a Stormy Sky*
mid 1820s (fig. 91)
Watercolour 36.2 × 58.4 (14¹⁄₄ × 23)
TB CXCVI Y; D17189

67 *Firing Rockets off the Coast* mid 1820s
(fig. 89)
Watercolour 22 × 28.8 (8⁵⁄₈ × 11³⁄₈)
TB CCCLXIV 134; D35977

68 *Life Boat and Manby Apparatus Going
off to a Stranded Vessel Making Signal
(Blue Lights) of Distress* 1831 (fig. 86)
Oil on canvas 91.4 × 122 (36 × 48)
The Board of Trustees of the
Victoria and Albert Museum
*Exhibited at the Victoria and Albert
Museum*

69 *?Ehrenbreitstein from Coblenz* c.1841
(fig. 78)
Watercolour 24.3 × 34.3
(9¹⁄₂ × 13¹⁄₂)
TB CCCLXIV 284; D36137

70 *Honfleur* 1830s (fig. 79)
Gouache and watercolour 13.6 × 19
(5³⁄₈ × 7¹⁄₂)
TB CCLIX 75; D24640

James B. Allen,
after J.M.W. Turner

71 *Caudebec* 1834 (fig. 80)
Engraving 9.6 × 13.8 (3³⁄₄ × 5³⁄₈)
Tate Gallery. Transferred from the
Library 1989
R462; T05601

72 *Steamer Leaving Harbour* mid 1840s
(fig. 73)
from dismembered *Whalers*
sketchbook
Charcoal and grey wash 22.2 × 33
(8³⁄₄ × 13)
TB CCCLIII 5; D35244

73 *Steamer with Yellow Smoke, Margate*
1840s (fig. 85)
Watercolour 24.1 × 29.2
(9½ × 11½)
TB CCCLXIV 173; D36017

Robert Carrick,
after J.M.W. Turner
74 *Rockets and Blue Lights* 1852 (fig. 82)
Chromolithograph on paper,
backed onto board 21½ × 29¼
(54.7 × 74.4)
The Board of Trustees of the
Victoria and Albert Museum

SKETCHBOOKS

75 *Falmouth Bay* 1811 (fig. 71)
Ivybridge to Penzance sketchbook
Pencil 16.8 × 21.8 (6⅝ × 8½)
TB CXXV, p.24a; D08901

76 *River Scene with a Smoky Steamboat*
c.1815–17 (fig. 72)
Walmer Ferry sketchbook
Pencil and white chalk 16 × 11.4
(6¼ × 4½)
TB CXLII, p.81; D10746

Captain George William Manby
(Designer)
77 *Manby apparatus mortar* (fig. 87)
Wood and iron 25 × 20 × 40
(9⅞ × 7⅞ × 15¾)
King's Lynn Museums, Norfolk
Museums Service

6 Industry and Construction after Waterloo

78 *Crossing the Brook* exh. 1815 (fig. 92)†
Oil on canvas 193 × 165.1 (76 × 65)
Tate Gallery. Bequeathed by the
artist 1856
N00497

79 *Leeds* 1815–16 (fig. 94)
Devonshire Rivers No.3 and Wharfedale
sketchbook
Pencil 17.8 × 26.7 (7 × 10½)
TB CXXXIV, pp. 79–80; D09883–4

80 *'The Hurries' – Coal Boats Loading,
North Shields* ?1822 (fig. 95)
Pencil 22.2 × 35.6 (8¾ × 14)
Yale Center for British Art, Paul
Mellon Collection

81 *Shields, on the River Tyne* 1823
(fig. 96)
Watercolour 15.4 × 21.6 (6 × 8½)
TB CCVIII v; D18155

82 *Kirkstall Lock, on the River Aire* 1824–5
(fig. 98)
Watercolour 16 × 23.5 (6¼ × 9¼)
TB CCVIII L; D18145

83 *A Vaulted Hall – perhaps 'Banquet
in the Thames Tunnel, 10th November
1827'* (fig. 103)†
[also known as *A Vaulted Hall* or
The Long Cellar at Petworth]
Oil on wood 75 × 91.5 (29½ × 36)
Tate Gallery. Bequeathed by the
artist 1856
T05539

84 *Chichester Canal* c.1828 (fig. 99)†
Oil on canvas 65.5 × 134.5
(25¾ × 53)
Tate Gallery. Bequeathed by the
artist 1856
N00560

85 *The Chain Pier, Brighton* c.1828
(fig. 100)†
Oil on canvas 71.1 × 136.5
(28 × 53¾)
Tate Gallery. Bequeathed by the
artist 1856
N02064

86 *Dudley, Worcestershire* c.1832
(fig. 101)
Watercolour and bodycolour
29 × 43.1 (11⅜ × 16⅞)
Board of Trustees of the National
Museums & Galleries on
Merseyside
(Lady Lever Art Gallery, Port
Sunlight)

87 *The Thames above Waterloo Bridge
('Waterloo Bridge – The Procession
before the Opening of London Bridge,
1st August 1831')* c.1832
(fig. 106)
Oil on canvas 90.5 × 121
(35⅝ × 47⅝)
Tate Gallery. Bequeathed by the
artist 1856
N01992

88 *Freiburg with Suspension Bridge* 1842
(fig. 110)
from dismembered *Freiburg*
sketchbook
Pencil and red ink 23.3 × 33.5
(9⅛ × 13⅛)
TB CCCXXXV 8; D33548

89 *Rain, Steam and Speed – the Great
Western Railway* exh. 1844
(fig. 104)
Oil on canvas 91 × 122 (35⅞ × 48)
National Gallery, London

Isambard Kingdom Brunel
90 *Arch Analyses for Maidenhead Bridge*,
1837, General Calculation Book,
1834–41, pp. 7–8 (fig. 105)
Ink on paper bound in cloth
32.6 × 20.6 × 3.2 (12⅞ × 8⅛ × 1¼)
By permission of the Librarian of
the University of Bristol

91 *Burning Blubber* mid 1840s
(fig. 113)
from dismembered *Whalers*
sketchbook
Chalk and wash 22.2 × 33
(8¾ × 13)
TB CCCLIII 6; D35245

92 *Burning Blubber* mid 1840s
(fig. 114)
from dismembered *Whalers*
sketchbook
Chalk and wash 22.2 × 33
(8¾ × 13)
TB CCCLIII 7; D35246

93 *'Hurrah! for the Whaler Erebus! Another
Fish!'* exh. 1846 (fig. 111)†
Oil on canvas 90 × 121
(35⅜ × 47⅝)
Tate Gallery. Bequeathed by the
artist 1856
N00546

94 *Whalers (Boiling Blubber) Entangled in
Flaw Ice, Endeavouring to Extricate
Themselves* exh. 1846 (fig. 112)†
Oil on canvas 90 × 120
(35⅜ × 47¼)
Tate Gallery. Bequeathed by the
artist 1856
N00547

95 *The Hero of a Hundred Fights*
*c.*1806–7, reworked and exh. 1847
(fig. 115)
Oil on canvas 91 × 121 (35⅞ ×
47⅝)
Tate Gallery. Bequeathed by the
artist 1856
N00551

SKETCHBOOKS

96 *London Bridge under Construction*
*c.*1827 (figs. 107, 108)
Isle of Wight sketchbook
Pencil 11.1 × 18.8 (4⅜ × 7⅜)
TB CCXXVII, pp. 12, 32a; D20751,
D20784

97 *Birmingham – Landscape Studies* 1830
(fig. 102)
Kenilworth sketchbook
Pencil 11.3 × 19 (4½ × 7½)
TB CCXXXVIII, pp. 18a–19*;
D22007–8

98 *Studies of Industrial Landscapes*
1830
Birmingham & Coventry
sketchbook
Pencil 6.7 × 10.7 (2⅝ × 4¼)
TB CCXL, p. 27; D22371

7 The Living Earth

W.B. Cooke,
after J.M.W. Turner
99 *Lulworth Cove, Dorsetshire* 1814
(fig. 119)
from *Picturesque Views on the
Southern Coast of England*
Engraving 14.6 × 21.7
(5¾ × 8½)
Tate Gallery. Purchased 1988
R92; T05348

100 *Eruption of the Souffrier Mountains,
in the Island of St Vincent, at
Midnight, on the 30th April, 1812,
from a Sketch Taken at the Time by
Hugh P. Keane, Esq. 1815*
(fig. 126)
Oil on canvas 81 × 106
(31⅞ × 41¾)
University of Liverpool Art
Collections

Robert Wallis,
after J.M.W. Turner
101 *Stone Henge, Wiltshire* 1829
(fig. 118)
from *Picturesque Views in England
and Wales*
Engraving 16.6 × 23.4
(6½ × 9¼)
Tate Gallery. Purchased 1986
R235; T04549

102 *Sketch for 'Ulysses Deriding
Polyphemus'* ?1828 (fig. 127)
Oil on canvas 60 × 89.2
(23⅝ × 35)
Tate Gallery. Bequeathed by the
artist 1856
N02958
On display from May 1998

W. Miller,
after J.M.W. Turner
103 *Glencoe c.*1834–6 (fig. 124)
Engraving 15.1 × 21 (6 × 8¼)
R549; T04989

104 *Snow Storm – Steam-boat off a
Harbour's Mouth Making Signals in
Shallow Water, and Going by the
Lead. The Author was in this Storm on
the Night the Ariel left Harwich* 1842
(fig. 129)
Oil on canvas 91.5 × 122 (36 × 48)
Tate Gallery. Bequeathed by the
artist 1856
N00530

SKETCHBOOKS

105 *Geological Formations at Lulworth
Cove* 1811 (fig. 120)
Corfe to Dartmouth sketchbook
Pencil 16.8 × 20.3 (6⅝ × 8)
TB CXXIV, p. 22; D08833

106 *Close-Up Views of Rock Faces* 1811
(fig. 121)
Stonehenge sketchbook
Pencil 22.7 × 18.5 (9 × 7¼)
TB CXXV(b), pp. 7–8; D41380–1
(not drawn at Stonehenge, unlike
later pages in sketchbook, but in
Devon)

107 *Interior of Fingal's Cave, off the Isle
of Staffa* 1831 (fig. 122)
Staffa sketchbook
Pencil 18.7 × 11.6 (7⅜ × 4⅝)
TB CCLXXIII, pp. 28–29*;
D26795–6

Edward Goodall,
after J.M.W. Turner
108a *Fingal's Cave* 1834 (fig. 125a)
title page from Sir Walter
Scott's *Poetical Works*, vol. X 1834
Engraving 16.5 × 11
(6¼ × 4⅜) (book size)
Dr Jan Piggott

Henry Le Keux,
after J.M.W. Turner
108b *Loch Coriskin* 1834 (fig. 125b)
facing title page from Sir Walter
Scott's *Poetical Works*, vol. X 1834
Engraving 16.5 × 11 (6¼ × 4⅜)
(book size)
Dr Jan Piggott

Michael Faraday
109 *Magnetic Field Experiment* 1851
(fig. 128)
Iron filings 59.2 × 42.6
(23¼ × 16¾)
Royal Institution of Great
Britain

INDEX

Adam brothers 29
Admiralty Screen and Building… 30; fig. 22
Aegina, Temple of Panellenius 32–3
Airy, George Biddle 128
Allason, Thomas 33
Allen, James B.
　Caudebec (after Turner) 79; fig. 80
Analytical Society 17
Anaxagoras 61
Ancient Italy 91
Angel Standing in the Sun, The 113
Annual Tours – Wanderings by the Seine 83
Apullia in Search of Appulus 36
architecture 21–36
Artists' Colourman's Workshop, An 52; fig. 44
Arundel Castle, with Rainbow 65; fig. 66
Ashburnham 53–4; fig. 48
astronomy 15, 17, 53, 58–9, 61–3, 68–72
Athenaeum Club 17, 100, 107–8

Babbage, Charles 15, 19, 126; fig. 6
Bacon, Anthony, the Elder 43, 45
Bacon, Anthony, the Younger 45
balloon flights 72
Banks, Sir Joseph 15
Banquet in the Thames Tunnel… see *A Vaulted Hall*
Basle and Bridge 68
Beckford, William 24–5
Beechey, Sir William 12
Bell Rock Lighthouse, The 90–1; fig. 90
Berger, J.F. 118–19
Between Quilleboeuf and Villequier 79, 83; fig. 77
Bicknell, Elhanan 106–7, 109–10
Birmingham 47, 97
Birmingham & Coventry sketch-book fig. 103
Birmingham – Landscape Studies 97; fig. 102
Blacksmith's Shop 49; fig. 43
Blaikley, Alexander 127
Blizard, William 12
Booth, Sophia 20, 26, 84
botany 15
Bournon, Count Jacques-Louis 12

Brentford 21, 37
Brewster, Sir David 15, 20, 65
bridges 34, 36, 92, 102–5; figs. 26, 27, 92, 105, 107, 110
Bridport, Dorsetshire 117
Brightling Observatory 53, 63; fig. 59
Brighton from the Sea see *Chain Pier, Brighton, The*
Bristol 23, 37, 38, 47
Broderip, William 15, 108–9
Brownrigg, William 43, 45
Brunel, Isambard Kingdom 98, 100, 102–3; fig. 105
Brunel, Sir Mark Isambard 98
Buchan, David 126
Buckland, William 14, 115, 125, 127; fig. 5
Burke, Edmund 23
Burning Blubber (cat. 91) fig. 113
Burning Blubber (cat. 92) fig. 114
Burning of the Houses of Lords and Commons, The 103–4
Burton, Decimus 112
Byron, George Gordon, 6th Baron Byron 66

Caledonian Comet 61
Callcott, Augustus Wall 11, 100
Calm Morning 30
Calstock Bridge 92
camera lucida 19
Campbell, Thomas 14, 83, 127–8
canals 9, 95–8
Canova, Antonio 66, 68
Carlisle, Sir Anthony 15
Carrick, Robert
　Rockets and Blue Lights (after Turner) 83; fig. 82
Cassiobury Park 49, 51
Caudebec 79; fig. 80
Caus, Salomon de 32
Chain Pier, Brighton, The 83–4, 96–7; fig. 100
Chambers, Sir William 12, 23, 29
Chantrey, Sir Francis 11, 17, 65, 72, 115, 117, 126
　Bust of John Fuller MP fig. 3
　Bust of Mary Somerville fig. 7
chemistry 10, 14, 16, 68, 127
Chemistry and 'Apuleia' sketchbook 10, 52
Chichester Canal 96 7; fig. 99
Cicero at his Villa 72
Clark, William Tierney 36
Claude Lorrain 47, 52, 92

Claudet, Antoine
　Charles Babbage fig. 6
Close-Up View of Rock Face fig. 121
Coalbrookdale 42–3, 53; fig. 36
Cockerell, C.R. 32–3, 126
colour making 10, 52; fig. 51
Constable, John 14, 64
　Waterloo Bridge from Whitehall Stairs, June 18th 1817 104
Cooke, George 55
Cooke, W.B. 55
　Lulworth Cove, Dorsetshire (after Turner) fig. 119
　Vale of Heathfield (after Turner) fig. 49
Cornwall 54–5, 92
Cottages and a Windmill 39–40; fig. 32
Country Blacksmith Disputing upon the Price of Iron …, A 49, 52; fig. 42
Covent Garden 13, 21, 23, 37
Coventry 47, 65, 97
Coventry 65
Cozens, Alexander 58
Cozens, John Robert 58
Crawshay, Richard 43, 45
Crossing the Brook 36, 92; fig. 92
Crowhurst 54, 92; fig. 50
Crystal Palace 9, 36
Cunn, Samuel 32
Cuyp, Aelbert 52
Cyfarthfa Iron Works 43, 45–6, 51, 54; figs. 38, 39

daguerreotypes 20
Danby, Hannah 20
Daniell, Thomas and William 12
Dartmoor 92
Dartmouth on the River Dart 76; fig. 74
Darwin, Erasmus 125–6
Davies, J. 34
Davy, Sir Humphry 14, 16, 17, 19, 65–8, 72, 76, 113–14, 126; fig. 4
Derbyshire 37
Devon 54–5, 92
Devonshire Coast No. 1 sketchbook 118
Devonshire Rivers No. 3 and Wharfdale sketchbook fig. 94
Diagrammatic Progress of the Setting Sun fig. 57
Dickinson, Thomas
　Narrative of the Operations for the Recovery of the …Treasure 115

Dido and Aeneas 10; fig. 1
Dido Building Carthage 34
Difference Engine 19
Donaldson, Thomas L. 33, 36, 76
Dorchester Mead, Oxfordshire 52–3; figs. 46, 47
Dover 37, 78
Dover 78; fig. 75
Dover (engraving) 78; fig. 76
Doyle, John fig. 13
Dudley, Worcestershire 97–8; fig. 101
Durham, North Shore sketchbook 34; fig. 26

Eagles, Revd John 91
Eastlake, Charles 11
Eastlake, Lady 51
Eclipse sketchbook 59; fig. 55
Eclipse, with a Gesticulating figure 61, 63; fig. 58
eclipses 58–9, 61, 63; figs. 55, 58
Eddystone Light House, The 90–1
Edgeworth, Maria 16
Edinburgh 65, 92
Edridge, Henry 12, 51
Egremont, 3rd Earl of 14, 96–7; fig. 2
?Ehrenbreitstein from Coblenz 79; fig. 78
electricity 16
Englefield, Henry 12
Entablature of a Tuscan Column … 30; fig. 24
Eruption of the Souffrier Mountains, The 121, 124; fig. 126
Essex, Lord 49, 51
Etna 125
Euclid 32, 119
Evening of the Deluge, The 108
Evesham 23

'Fallacies of Hope, The' 72, 97
Falmouth Bay fig. 71
Faraday, Michael 16, 19, 89, 126, 128; figs. 8, 13, 128
Faringdon, Joseph 51, 67
Farnley Hall 29, 64
Fawkes, Hawksworth 36
Fawkes, Walter 29, 92, 115
Fighting Temeraire …, The 74, 91
Finch, Robert 36
Fingal's Cave 120–1; fig. 125a
Firing Rockets off the Coast fig. 87
Fishing Boats with Hucksters Bargaining for Fish 84
Folkstone sketchbook 91
Fonthill 24–5; figs. 17, 18

Fonthill sketchbook 25; fig. 18
Forge Scene 10, 46, 47; fig. 40
foundries and forges *see* iron-working
Foundry with Tilt Hammer 47; fig. 41
Fountain of Fallacy, The ('The Fountain of Indolence') 71–2; fig. 70
Freiburg sketchbook 36; fig. 110
Freiburg with Suspension Bridge 36; fig. 110
Fuller, John 14, 53–4, 63; fig. 3

Galileo Galilei 69
Galileo's Villa 68–69; fig. 67
Geological Society 17, 118–19, 121
geology 11, 14–15, 16, 17, 115–28
Gilpin, William 23, 39
Girtin, Thomas 46
Glencoe 120; figs. 123, 124
Goats 68
Goodall, Edward
    Fingal's Cave (after Turner) fig. 125a
Gray, J.E. 107, 110
Great Western Railway 102–3; fig. 105
Grecian manner 29
Green, Charles 72
Greenwich Hospital 63; fig. 60
Griffiths, Thomas 91

Hakewill, James 33
Hamilton, John 32
Hampshire 37
Harbour with figures and Shipping, Margate 84; fig. 84
Hardwick, Philip 33
Hardwick, Thomas, the Elder 21
Hardwick, Thomas, the Younger 21, 23, 29, 33
Hart, Solomon
    Galileo, when Imprisoned ... 72
Harvest Home 49, 51
Harvest Home sketchbook 49
Hastings sketchbook fig. 59
Hatchett, Charles 12
Hawksmoor, Nicholas 33
Haydocke, R. 32
Haydon, Benjamin Robert 126
Hearne, Thomas 12, 46, 51
Hereford 37
Hereford Court sketchbook 45
Hero of a Hundred Fights, The 9–10, 49, 51, 111–14; fig. 115
Herschel, John 15, 19
Herschel, Sir William 15
Hesperides II sketchbook fig. 47
High Street, Oxford 30
Highmore, Joseph 32
Hoare, Sir Richard Colt 42
Holland, Henry 12, 23
Hollond, Robert 72
Holworthy, James 33–4
Honfleur 79; fig. 79
(?)*Honiton Mill* 39

Hooker, Joseph Dalton 107–8, 110
Hooker, Sir William 15
Horsburgh, John 90
Horsley, J.C. 100
Hullmandel, Charles 19, 127
Hurrah! for the Whaler Erebus! ... 106; fig. 111
'Hurries, The' ... 92, 95; fig. 95
Hutton, James 125

Imaginative Sunset fig. 69
industrial subjects 37–57, 92–8
Interior of fingal's Cave ... fig. 122
Interior of a Prison 30; fig. 23
Interior of a Workshop 52; fig. 45
Iron Foundry, An 9, 40, 47; fig. 34
ironworking 9, 37, 39–40, 42–3, 45–9, 53, 4, 98, 111–12; figs. 36–43
Isle of Wight 37
Isle of Wight sketchbook figs. 33, 108
Ivybridge to Penzance sketchbook fig. 71

Jackson, John 12, 65
Jones, George 117
    The Royal Procession at the Opening of London Bridge ... 105; fig. 109
Juliet and her Nurse 72, 103

kaleidoscope 15
Keane, Hugh P. 121
Keelmen Heaving in Coals by Moonlight 95, 106; fig. 97
Kenilworth 97
King's Visit to Scotland sketchbook 92; fig. 56
Kirby, Joshua 32
    Dr Brook Taylor's Method of Perspective Made Easy 21
Kirkstall Lock, on the River Aire 36, 95–6; fig. 98

Landscape with Workmen Building a Bridge 34; fig. 26
Laporte, John 12
Laurenstein 36; fig. 28
Lawrence, Sir Thomas 11, 17, 65, 67, 72
Le Keux, Henry
    Loch Coriskin (after Turner) fig. 125b
Leeds 92; figs. 93, 94
Lenox, James 79
Leslie, C.R. 100
Lewis, F.C.
    Coalbrook Dale (engraving after Turner) 43
Liber Studiorum 49, 54, 68, 117; fig. 50
Life Class sketchbook fig. 57
Life-Boat and Manby Apparatus ... 87–90, 102; fig. 86
lighthouses 84, 90–1; fig. 90
lime-burning 42–3

Limekiln at Coalbrookdale 42–3, 92, 95, 97–8; fig. 36
Lloyd, Hannibal Evans 98
Loch Coriskin 120, 125; fig. 125b
Lomazzo, Giovanni Paolo 32
London 30
London, building in 21, 23, 34, 36
London Bridge 105
London Bridge under Construction figs. 107–8
Long Acre 37
Longships Lighthouse, Land's End 91
Loutherbourg, Philippe-Jacques de 40, 42, 43
    Welsh & Shropshire Industrial Subjects ... fig. 37
low-life subjects 49, 51–2, 55
Lowther sketchbook 61
Lubbock, J.W. 109
Lulworth Cove, Dorsetshire 119; fig. 119
Lupton, T. 90
Lyell, Charles 14, 115

MacCulloch, John 14, 115, 117–19, 121
magnetism 16, 19, 20, 126–8; fig. 128
Maidenhead Bridge 102–3; fig. 105
Malton, Thomas 21, 32
Manby Apparatus 87, 89–90, 102; figs. 86–8
Manby, Captain George 14, 84, 87–90, 102
Marford Mill, Denbighshire 39; fig. 30
Margate 21, 46–7, 84, 128; figs. 84, 85
Martin, John 63
Mason, Monck 72
mathematics 15, 17
Mayall, J.J.E. 10, 20, 126
meteorology 63–8
Mew Stone, The 117
Midlands 37, 97
mills 37, 39–40, 48, 92; figs. 30–2
Milton, John 69, 71
Minehead 117
Modern Italy 91
Monro, Dr Thomas 43, 58
Monte, Guidobaldo del 30
Moore, Harriet 19, 126–7; fig. 9
Moore, James Carrick 20, 126
Moore, Thomas 65–6
Morning after the Deluge, The 108
Mottram, Charles
    Samuel Rogers at his Breakfast Table (after Doyle) fig. 13
Moxon, Joseph 32
Murray II, John 20
Murray III, John 20
Mustering of the Warrior Angels, The 69; fig. 68

Naples 67

Napoleonic Wars 9, 45, 92
Nash, John 33
Nasmyth, Alexander 68
natural history 15, 16
Natural History Museum 109
Natural Philosophy 12
Nelson sketchbook 95
New Moon, The 84
Nixon, Revd Robert 12
Noble, Matthew
    Bust of Michael Faraday fig. 8
North Devon sketchbook 55
Notes on Architectural Perspective fig. 25
Nugent, Nicholas 121
Nunwell and Brading from Bembridge Mill 40; fig. 33

Opening of the Vintage of Macon, The 36
Opening of the Wallhalla, The 108
optics 15
Orfordness 91
Owen, Sir Richard 11, 15, 19, 108–10; fig. 10
Oxford 21, 23–4, 115
Oxford Almanack 23–4
Oxford sketchbook 21; fig. 15

Pamplin, Richard 46
Pantheon, the Morning after the Fire, The 38
Pantheon Opera House 37
Paris, Seine and Dieppe sketchbook 29
Parry, Sir Edward 126
Part of the South Front of Fonthill Abbey 25; fig. 18
Pass of St Gothard, The 115; fig. 117
Payne Knight, Richard 23
Peace 91
Peace, Burial at Sea 83
Pearson, Dr George 12
Pembury Mill, Kent 39
people at work, depictions of 38–9, 48
perspective 21, 23, 26, 30, 32, 33, 72, 119; figs. 15, 22–4
Perspective sketchbook 30; fig. 25
Pettigrew, Thomas 20
Pevensey Bay from Crowhurst Park 54
Phillips, Thomas 11, 14, 68, 76
    3rd Earl of Egremont fig. 2
    Sir Humphry Davy fig. 4
    William Buckland fig. 5
photography 10, 20
Picturesque movement 9, 23, 39
Picturesque Tour of Italy, A 33
Picturesque Views of the Antiquities of Pola, in Istria 33
Picturesque Views of England and Wales 46, 54, 65, 78, 97; fig. 118
Picturesque Views on the Southern Coast of England 119; fig. 119
Pound, Daniel J. 20

Poussin, Nicolas 47
Price, Uvedale 23
Priestley, Joseph 32
Projected Design for Fonthill Abbey, Wiltshire, A (with Wyatt) 24; fig. 17
Pye, John 64

Queen Anne Street house 10, 20, 25–6, 29, 59

railways 9, 95, 102-3; fig. 105
Rain, Steam, and Speed ... 36, 102–3; fig. 105
Rainbow with Cattle, A 65; fig. 65
rainbows 65; figs. 65, 66
Raphael 67, 68
Reaping 49
Rembrandt van Rijn 19, 42–3
 Landscape with the Rest on the Flight into Egypt 42; fig. 35
Rennie, John 84
Rising Squall, The: Hot Wells from St Vincent's Rock, Bristol 38
River sketchbook 49
River Margate sketchbook fig. 45
River Scene with a Smoky Steamboat fig. 72
Rivers of England, The 92, 96; figs. 74, 96, 98
Rivers of Europe, The fig. 63
roads 9, 95–6
Roberts, David 76
Rockets and Blue Lights ... 83, 91; fig. 82
Rogers, Samuel 20, 63, 69, 71; fig. 13
Romanticism 115
Rome 17, 19, 36, 65–8
Rome, from Mount Aventine 72
'*Rome, from the Vatican*', *Study for* 36; fig. 29
Rome from the Vatican. Raffaelle, Accompanied by La Fornarina ... 36, 68
Rosehill 53
Ross, James Clark 107, 110, 126
Rosse, Earl of 19
Royal Academy 9–10, 12, 21, 23, 24, 29–30, 47, 67, 92, 100, 111
 Professor of Perspective 26, 30, 32, 33, 119; figs. 22–4
Royal College of Surgeons 19, 108–9
Royal Institution 13–14, 16, 53, 89, 127
Royal Society 12, 19, 68, 90, 100, 109, 126, 128
Rumford, Sir Benjamin Thompson, Count 12, 13–14
Ruskin, John 11, 46, 68, 74, 76, 91, 115

sailing vessels 74, 84, 91

St Mary's and the Radcliffe Camera from Oriel Lane, Oxford 23–4; fig. 16
Salt Hill, Slough 64
Sandby, Paul 21, 46
Sandby, Thomas 21, 23
Sandycombe Lodge 26, 28–30, 34; figs. 19, 20
Sandycombe and Yorkshire sketchbook 26, 28; fig. 19
Scarborough II sketchbook 63; fig. 61
Schematic Memorandum of Sunrise fig. 56
Schiller, Friedrich von
 Song of the Bell 113–14
Scotland 48, 65, 120–1
Scotland and London sketchbook 34; fig. 27
Scott, Sir Walter 81–2, 120–1
Sedgwick, Adam 14
Shadrach, Meshech and Abednego ... 105
Shields, on the River Tyne 92, 98, 106; fig. 96
Shipley, William 13
shipwrecks 74
Siringatti, Lorenzo 32
Skene, Revd James 65
skies 58–73
Skies sketchbook 63–4; fig. 62
Skyscape over Hawksworth Moor, with Colour Notes 63; fig. 61
Slavers Throwing Overboard the Dead and Dying ... 91
Small High Cloud Catching the Sunset (Sunset at Sea) fig. 64
'Smirke, Mr' 12
Smirke, Sir Robert 53
Snow Storm – Steam-Boat off a Harbour's Mouth ... 82–3, 128; fig. 129
Soane, Sir John 12, 20, 28, 32
Society of Antiquaries 12
Society for the Encouragement of Arts, Manufactures and Commerce 12–13, 37–8
Somerset House 12, 23, 90, 105
Somerville, Mary 11, 15–16, 19–20, 68–9, 71–2, 124, 126, 128; fig. 7
 Italian Landscape: View towards a Bay fig. 12
 Mechanism of the Heavens 68–9, 71
 Mole de Gaeta fig. 11
 On the Connexion of the Physical Sciences 16, 17, 71–2, 128
 Physical Geography 16
Somerville, Dr William 19, 68, 128
Staffa sketchbook fig. 122
Staffa, Fingal's Cave 79, 81, 119; fig. 81

stage machinery 37
Stanfield, Clarkson 72, 100
steam power 39, 74–84, 91, 97, 104–5; figs. 72–4
Steamer Leaving Harbour 84; fig. 73
Steamer with Yellow Smoke, Margate 84; fig. 85
Stevenson, Robert 90–1
Stokes, Charles 115, 117–18, 126
Stonehenge sketchbook fig. 121
Stone Henge, Wiltshire 117–18; fig. 118
Study of Fortifications at Laurenstein 36; fig. 28
Studies inside a Picture Gallery fig. 21
Studies for Vignettes sketchbook fig. 67
Study of a Partial Eclipse of the Sun 58; fig. 55
Study of the Sky 63–4; fig. 62
Sublime, The 39
Sunset – Nancy ... fig. 63
Swans sketchbook 46, 47; fig. 41

Tabley No. 3 sketchbook 29; fig. 21
Taylor, Brook 32
Thames above Waterloo Bridge, The 103–5, 106; fig. 106
Thames Tunnel 98, 100, 102; fig. 104
Thomson, James
 The Castle of Indolence 71
Thornbury, Walter 21, 65
Tintagel Castle 55; fig. 52
Tintagel Castle 1818 55; fig. 53
Tintagel Castle, with Lifting Gear 55; fig. 51
Tivoli and Rome sketchbook 68
Trimmer, Henry Scott 21, 59, 61
Trimmer, Henry Syer 59, 61
Trimmer, James 21
Trimmer, Sarah 21
Trout Fishing in the Dee 51
Turner, Charles 121
Turner, Dawson 76, 89, 107–8, 115
Tyndall, John 127

Ulysses Deriding Polyphemus 125–6; fig. 127
Unpaid Bill, The 51, 52

Vale of Ashburnham 53–4, 92; fig. 48
Vale of Heathfield 54, 92; fig. 49
Vanbrugh, Sir John 33
Varley, John 12
Vaulted Hall, A 98, 100, 102; fig. 103
Vesuvius in Eruption 121, 124–5
vulcanism 14, 121, 124–5; figs. 126, 125

Wagmüller, Michael
 Bust of Sir Richard Owen fig. 10
Wales 37, 43, 45–6, 47
Wallace, Robert K. 106–7
Wallis, Robert 98
 Stone Henge, Wiltshire (after Turner) fig. 118
Walmer Ferry sketchbook fig. 72
Wanstead New Church 21, 23; fig. 14
War 91
Warwick 97
Waterloo Bridge from Beneath 34; fig. 27
Waterloo Bridge – The Procession before the Opening ..., 103–5, 106; fig. 106
Waterloo and Rhine sketchbook 36; fig. 28
Watermill and Stream 39; fig. 31
watermills *see* mills
Wellington, Duke of 10, 111–14
Wells, W.F. 115
Wethered, William 91
Whalers sketchbook figs. 73, 113, 114
Whalers (Boiling Blubber) ... 106; fig. 112
whaling subjects 106–10; figs. 111–14
Whewell, William 14
Wilkie, Sir David 19, 72
 The Village Politicians 48–9
Willmore, J.T.
 Dover (after Turner) fig. 76
Wilson sketchbook 9, 46; fig. 40
Wilson, Richard 46
Windmill and Lock sketchbook 26, 28
windmills *see* mills
Windsor Castle 64
Windy Day 30
Wollaston, William 126
Woodcock Shooting sketchbook 26
Worcester 97
Wreckers – Coast of Northumberland ... 83–4; fig. 83
Wright of Derby, Joseph 40, 42, 121
 An Iron Forge 42
Wyatt, James 24–5, 32
 A Projected Design for Fonthill Abbey, Wiltshire (with Turner) 24; fig. 17
Wyatt, Matthew Cotes
 Wellington Military Memorial 111–14; fig. 116
Wyatt (Wyatville), Jeffry 32

Zoological Society 17
zoology 15, 17, 107–9